Numerical Intuition: Familiarity

Michael Butler

Copyright Notice

ISBN

ISBN: 978-1-7636544-1-9

Legal Disclaimer

Trademarks

Library and Archives Australia Cataloguing in Publication

A catalogue record for this book is available from the National Library of Australia.

First Edition

Printed in Australia

Publisher Information

Published by:
ECS PUBLICATIONS PTY LTD
Australia

Contact Information

For permissions and other inquiries, please contact:
ECS PUBLICATIONS PTY LTD
Email: ecs.publications@gmail.com

About the Author

Michael Butler is an independent writer, publishing texts on the topics of mathematics, history and philosophy. Born and brought up in Melbourne, Australia, Butler developed a keen interest in mathematics and history through high school and early university studies.

Through his time in mathematical and historical study, Butler has observed trends in the development of mathematics and key historical events, presenting essays on their relations and the paradigm shifts that prompted the discovery of key mathematical results.

Having studied mathematics and computer science at an undergraduate level at reputed Australian institutions, Butler has himself a strong understanding of foundational mathematical concepts and their pedagogical basis. With this understanding of contemporary mathematics, Butler presents compelling narratives on the methods of mathematical learning and the development of the mathematical field along with its present pedagogical methods.

When he is not writing, Butler enjoys time out in the wilderness and park life of Victoria, home to some of most well preserved natural terrain in the world including Ninety-Mile Beach and Wilson's Promontory. Butler is also actively involved in the cricketing community of his locality, playing games in local competitions and teams.

Butler currently lives in the outer suburbs of Melbourne, Australia with his family and is currently working on his second text in the series *Numerical Intuition*, with the upcoming title *Numerical Intuition: Understanding.* Readers can engage with Butler through his website and email michael.butler@gmail.com for any queries and general contact.

Contents

Key Definitions

Throughout this excerpt, I use a variety of terms which may have slight variances from their conventional use cases. This section is to clarify the definitions of a number of key words and phrases I use within the text.

Practitioner: Any individual that practices mathematics. This can be a school student, university professor or an avid learner of mathematics with no institutional background..

Computation/Operation: I use the term 'computation' and 'operation' very broadly in this text. Anytime someone performs a calculation with numbers or has an attempt on a problem in mathematics, I treat it as a 'computation' or 'operation'.

Axiom/Postulate: A statement or assertion from which we can make logical deductions.

Mathematical Object: A number, function, shape (anything really) that's just mathematical in nature.

Field: an established area of study or pursuit.

For example:

1) Mathematics is a field
2) Biology is a field

3) Humanities is a field
4) Music is a field

Area: A distinct subject matter within a field.

For example, in the text I use the following interchangeably:

1) Genetics is an area of biology
2) Trigonometry is an area of mathematics
3) Biochemistry is an area of chemistry

Purpose behind the essay

Learning in general, and especially the domain of mathematical learning is full of misnomers. How do I know this, the reader may ask? Well I actually studied tertiary mathematics and computer science, and I can tell you first hand how misinterpreted the field is. In fact the experiences I have had within tertiary institutions and the prevailing conceptions surrounding mathematical learning are so disparate, that I felt a *necessity* to actually write this text. Thus the level of disparateness between one's personal experiences and the narratives surrounding mathematical learning, must be quite substantial if an individual feels compelled to actually write a book on the topic of mathematical learning.

So what then are the prevailing arguments? They go something like this:

1) I'm not intelligent so I can't do maths.
2) Maths is for smart people.
3) Maths is hard (which it is but we will elaborate on this later).
4) I'm not a maths person (*who is at the start?)*
5) I did one maths unit and it was hard. Therefore I'm going to do an easy non-mathsy unit.

And you get the gist.

One point to note though is that there is a significant amount of contemporary literature these days that promotes the topics of mindset and the overcoming of obstacles. The reader may think that this text is simply 'riding the wave' of such literature and this would not be an unjust assessment. With the number of best sellers and business oriented literature all giving a common message of self-improvement, leadership and inherent capability (you can achieve anything) which, I must admit, can also be derived from the concepts within this text; the reader could absolutely think that I am just another writer trying to profit from this wave.

I have two objections to this argument:

1) The concepts I introduce, I aim to keep as unbiased as possible as separate from the current wave of literature and the various ideals it promotes.
2) I actually care when I see droves of fellow university and high school students be disheartened by their results within this very topic (mathematics).

Now the reader will be able to verify the first point through examination of this text, but the second is far more difficult to verify unless you were to speak to the writer in person. To that end, I hope that the reader can resonate with my second point, through the tone of writing and the various examples/footnotes, I give to prove the points I contend.

Within my own intellectual pursuits there are a number of insights which I have had which I present in this book, and much of what I contend has its basis in anecdotal experience and observation.

Through rigorous study and contemplation of the anecdotal experiences and its relation to mathematical learning, I present what I believe (and show) is a key component to mathematical learning: *numerical intuition,* and its related concept: *familiarity.*

As a customary note within such texts, I have provided examples of past and contemporary mathematicians to highlight the key points I am making. I have specifically focused on the aspects of the individuals which most relate to the ideas of this book - not to downplay their achievements within their individual careers.

Part 1: Pretext

Niels Bohr (1885 - 1962): "An expert is a person who has made all the mistakes that can be made in a very narrow field."

Introduction

It's pretty fair to say that mathematics (as a field of study) is in a troublesome state. Despite the ubiquity of products and services built on mathematical foundations, mathematics itself is one of the most misunderstood fields of study within schools, universities and professional work environments. Having been tainted from the earliest stages of learning as a means of performing repetitive arithmetic operations (addition, subtraction, multiplication etc), mathematical thinking and understanding has been completely misinterpreted, leaving the field with a common negative perception among the general public. With the advancement of the fields of computer science and IT, mathematics has made a slight comeback in recent years with there being a greater emphasis on its connection to the modern IT landscape.

This is highlighted in the development of artificial intelligence, neural networks and image recognition all of which make use of mathematics (mostly matrices to be more specific and ridiculous amounts of data). However, these are *applications* of developed mathematics which previously existed as theory. It's simple to find applications of theory within the real world should one look hard enough, but it's significantly more difficult to have an understanding of the theorems the applications are built upon.

Better yet, a formalised method for obtaining an understanding of any mathematical theory remains unknown. Every teacher (lecturers, school teachers, course instructors) seems to offer their own framework for how concepts within mathematics (and well anything really) should be taught[1]. As a whole mathematical learning remains a puzzle to most modern psychologists and self-help authors who have focused more on the development of soft skills in recent years such as communication, public speaking and mindset/mentality.

Thus, mathematical learning has essentially remained untouched by such professionals and for good reason. When studies are performed to unmask the puzzle of mathematical learning all they tend to demonstrate is how one can better perform on some kind of aptitude test or intelligence measure.. In terms of actual mathematical theory development or general computational ability, researchers are aware that their studies would have no findings that could actually be put into practice in institutions. Don't believe me? A quick search on Udemy or any course provider will give you the answer. Entering 'mathematical learning' into the search filter and one will find a plethora of videos with titles such as 'mathematical foundations of machine learning', 'foundations of data science and machine learning' and 'master the fundamentals of maths'. How many such videos actually teach how an understanding of a theory can be *developed*? How many practitioners actually understand the core principles that govern

[1] Many of the readers may be aware of the introduction of digital resources (laptops, digital whiteboards and then soon after IPads and other tablet devices) into the Australian Education System following the Digital Education Revolution (DER) program introduced by then Prime Minister, Kevin Rudd in 2008. Whoever thought that teenagers would use these devices for learning instead of playing games or using social media was clearly misguided. Since the DER was introduced in Australia there has been a concerning decline in overall literacy and numeracy. This has been measured through various standardised assessments such as the National Assessment Program - Literacy and Numeracy (NAPLAN)[1].

mathematical learning, from the involvement of the human mind to the distinctive aspects of mathematics as a field of study? (We will discover these principles in a later section).

In stark contrast to studies performed on mathematical learning, there have been an innumerable number of studies done on the areas of business and leadership. Almost every university professor it seems has published literature offering their own philosophy on what creates a good business or renown leader.

Since the demand for such literature in our current age is exceedingly high, if any recommendations proposed in these texts carry any logic or anecdotal support, there is a significant chance they can actually be implemented within industry. A classic example of this is Lean & Six Sigma which has had entire books written on it with certifications being offered in applying this *philosophy*. I say philosophy on purpose. What we are referencing when we say 'Lean & Six Sigma' is a philosophy on how to achieve the core *principles* of minimising waste and maximising customer value. We already know that any business will be effective if it's applying the *principles* of Lean & Six Sigma of minimising waste and maximising customer value. These *principles* will necessarily lead to better outcomes for the business. It doesn't matter if one changes what the *philosophy* is to apply these principles. People fail to realise that the reason the Japanese even adopted the Lean model for business is because it's essentially their native culture being applied in business. The

Japanese are known for their cultural emphasis on continuous improvement (Kaizen), respect (Hitzokuri) which manifests as a maximisation of customer value and the elimination of waste (Muda)[[1]]. It just so happened that their philosophy was highly logical in creating effective business processes for car manufacturing[2]. Thus we can see that any *philosophy* that focuses on the key *principles* governing any domain will be an effective *philosophy,* which we can then call a variety of different terms including *theorem, process, procedure* or *workflow* - you get the gist.

Furthermore, we see that in areas such as communication, public speaking and my personal favourite 'disruptive business'[3] - business cases present themselves almost immediately for trying new methods and performing greater research to find effective ways of working. With mathematics, however, there exists no such incentive beyond the desire of a teacher to help their students learn what may be on their next test or examination.

---

[2] As a side note, just consider if a business hypothetically had waste that was actually of use in creating customer value (similar to how the by-products of milk production can be used to create protein powders). Now if any philosophy is applied that focuses on reducing waste, would this be an effective philosophy? The philosophy here is to minimize waste but such a business would actually want to produce more waste!

[3] I personally did a business course once on this very topic. The whole course focused on three companies: Tesla, AirBnB and Uber. Hardly a proper course on how to make a 'disruptive business' if you can only find three examples in history.

Observation and Theory

Within science and mathematics there exists an intricate relationship between three key concepts: observation, theory and principle. Prior to the creation of any theoretical framework what *usually* occurs is that a practitioner will make some observation about the real world. A very simple example is the observation of the classic apple falling to the ground by Newton. From this *observation* Newton then created a *theory* to explain why this phenomena occurred. Now the theories can be absolutely anything - Newton could have said that the apple fell to the ground because it actually had a consciousness and desired to drop to the ground. What then defines whether the theory is correct or not will be if its predictions hold true in all observational cases. Normally within the sciences there will be stages of peer review or testing that aim to falsify the predicted theory and if it cannot be falsified it is generally accepted after enough rigorous testing. Though this framework is highly logical and effective, it is different to our approach within this book. What we will aim to do is prove the process behind mathematical learning using statements/facts which are universally accepted - *principles*. This may seem unusual, but we actually make use of principles all the time within the sciences - we simply omit the word ‘principle’ in general use cases such as ‘the principle of conservation of energy’ (becomes conservation of energy), ‘the principle of homeostasis’ (becomes homeostasis) or ‘the principle of natural selection’ (becomes natural selection) just to name a few.

With established principles we can create a framework for mathematical learning, primarily through *reasoning* rather than by regular scientific methods. One reason we will work in this manner, is that scientific methods can have many limitations in their performance including their duration, funding, area/geography and methodology. Another key consideration is that of general biases; methods usually implemented within an institution are dependent upon the perspectives of the institution's *directors and managing bodies*. So any new program of learning will normally require some kind of approval from *actual* human beings before commencing. Human beings, as the reader likely already knows, can have a variety of reasons for disapproving any endeavour which are completely unrelated to the endeavour itself. Everything from having had a bad day, having a personal grievance with a lecturer or personally giving minimal importance to a field of study could be genuine reasons a director or managing body may prevent the implementation of an otherwise very effective program of learning.

Thus to overcome any limitations presented to us by scientific inquiry and bias, we will instead *reason* purely from what we do understand about mathematical learning to devise a way of learning which is logical and coherent.

Axioms and Postulates

One question which may arise in the reader's mind is how we can actually devise a framework for mathematical learning based on established principles. To answer this we can look to university level mathematics and the key concepts one is taught at the tertiary level. Within mathematics itself one of the first things you learn is to reason from axioms, and to draw results *only* from the axioms.

Axioms themselves are statements which do not require proof as they are the foundational statement from which other deductions are made - and axioms can be anything. For example, I could say something (which seems ridiculous) like $1 + 1 = 5$. Now this does not work if you try and contextualise such an example within the real world, however we can create a system of mathematics where this is a foundational statement. From this foundational statement we could then see that

$$2 + 2 = 1 + 1 + 1 + 1 = 5 + 5 = 10$$

Thus within the system we just devised we have that $2 + 2$ is actually equal to 10. Obviously this seems ridiculous because it does not align with the real world, but the deduction that $2 + 2 = 10$, comes from our initial *axiom or assertion* that $1 + 1 = 5$, which was simply based on choice. We could choose anything to be our base statements from which to reason, and this is not as absurd as what it may seem. From the choice of axioms and principles many seemingly ridiculous theories have come to be accepted within the sciences (mostly physics):

1) The wave-particle duality of a subatomic particle.
2) Time dilation in Einstein's theory of special relativity
3) Parallel lines 'meeting' in non-Euclidean spaces[4].
4) The 11 spatial dimensions of string theory.
5) The existence of antimatter etc.

I encourage the reader to think for a moment - how could the existence of such postulates be deduced if we cannot instrumentally observe any of them? That too these theories are reasoned to a point where investors are willing to spend millions of dollars with the confidence of discovery in the various theoretical proposals. Given below is a list of accepted theorems that were made from this 'axiom based reasoning' that had significant investment (either directly or indirectly)[5] to confirm their deductions:

1) General Relativity: Was discovered when two scientists, Arthur Eddington (1882 - 1944) and Andrew Crommelin (1865 - 1939) went on two separate expeditions (Arthur to

[4] In non-Euclidean geometries, such as spherical geometry or hyperbolic geometry, the underlying space is curved rather than flat. On a curved surface, lines that appear parallel can converge or diverge depending on the curvature of the space. In spherical geometry, for example, lines of longitude on the surface of a sphere are parallel at the equator but converge at the poles. This is because the surface of a sphere is positively curved, meaning the space "bends back on itself." As a result, parallel lines on a sphere eventually meet.

[5] Here when I say 'directly' I mean financially through direct investment. Indirectly, is through any non-financial investment such as travelling to remote locations or the creation of specific instruments for measurement.

Principe, an island off the west coast of Africa, and Andrew to Sobral in Brazil) so they could photograph the position of the stars close to the Sun's perimeter before and after an eclipse[3].

2) Higgs Boson Particle: Was discovered from experiments performed at the Large Hadron Collider (LHC) and had a total projected cost of almost 3.5 billion euros at the time of construction (excluding upgrades and maintenance)[4].
3) Pulsars and neutron stars: Discovered by Jocelyn Bell Burnell (1943 -) and Antony Hewish (1924 - 1921) in 1967. Bell Burnell, a graduate student at the time, was analyzing data from a radio telescope at the Mullard Radio Astronomy Observatory in Cambridge[5].
4) Black holes: Though accepted through indirect observations, the first direct evidence came in 2019 when the Event Horizon Telescope (EHT), a global network of existing radio wave observatories took the infamous photo of the shadow of the black hole against the bright emission from the surrounding accretion disk[6] [6].
5) Gravitational Waves: Discovered through observations made at the Laser Interferometer Gravitational-Wave Observatory (LIGO) (built across two sites, Hanford and Livingston in America). Each site has an interferometer which consists of two perpendicular arms several kilometers in length in order

---

[6] The event horizon of a black hole is a theoretical boundary surrounding the black hole beyond which nothing, not even light, can escape. The term though, in my opinion is a bit misleading, since not actually a single point. There is not one 'event horizon' there are infinitely many points beyond which nothing can escape. This is because what we call the black hole is a singularity, or individual point, and the actual 'event horizon' is a boundary around the singularity which is determined by the Schwarzschild radius.

> to detect disturbances in length caused by gravitational waves - and yes you guessed it, LIGO receives ongoing funding from the National Science Foundation (NSF); it's most recent upgrade in 2015 (just before the first detection of gravitational waves) was just over $200 million USD[7].

It's unusual in these cases as the scientists involved were, in a way, looking for an observation or created extensive apparatus that could confirm the predicted observations. This works in the reverse of how science usually works, there are normally observations made and then a theory is created to explain the observations (albeit with an initial hypothesis). It is rare that a theorem is made *before* an observation! Well maybe you can do this if you have the backing of mathematical logic and indisputable postulates. It is from the postulates which the scientist initially asserts that such deductions are made, and if such a prediction is wrong or misaligned to the real world we can simply change the postulate.

Principle Aim

If it's true that the principles of mathematical learning elude modern psychologists and evade extensively funded research, then how is one meant to establish a framework for mathematical learning?

We know that, in an institutional context, the results of any research study require implementation to actually observe their outcome, and we know that very often it is the *philosophy* of the researcher which is being implemented which can have no basis in established *principle*[7]. Much like how philosophies are used to maximise principles for desired outcomes (as with the Lean & Six Sigma), we can ask ourselves firstly, do there exist any principles that exist for mathematical learning?

From such principles, would it be possible to create a potential theory or formalisation for the learning of mathematics?

The aim of this book is to introduce a concept which we will refer to as numerical intuition which in fact (as we shall find), does underlie how mathematics is learnt. We will go through its various components beginning with the first stage of *framing,* where we will discuss what numerical intuition actually is and its various components.

---

[7] Case in point is the theory of 'learning styles'. This is when your teacher claimed that different people have different preferred ways of learning the same thing: auditory, visual and kinesthetic. Unusual how a disproportionate number of people had 'visual' as their preferred learning method (maybe because it requires the least amount of effort out of the three styles).

The second stage is referred to as *familiarity* where we will delve deep into what familiarity actually is and its relation to numerical intuition.

Following the stage of familiarity we will look at the third stage of *properties*, where we will analyse the distinctive properties of mathematics as a field and its relation to numerical intuition and the building of *familiarity.*

All throughout our discussion on the numerous aspects of numerical intuition and mathematics we will refer to mathematicians and scientists whose life experiences represent the case in point. This is to put the ideas discussed within this book into connection with real world cases.

Part 2: Frame

John Von Neumann (1903 - 1957): "In mathematics you don't understand things, you just get used to them."

Treatise on mathematical education

In the established structure that school systems are currently framed within the Western world, mathematical learning is very one dimensional (pun intended). What normally follows is the following sequence of learning: a student goes through endless hours of computation to solve questions all with the intention of improving their basic math skills such as addition, multiplication, integration etc. Then more complex subjects such as algebra and calculus are introduced where the same process of endless computation is repeated. Then if they feel they excel in these areas they might go on to study mathematics at university where they realise that everything they were taught at high school took hundreds of years of reasoning and development to arrive at.

Most people don't realise that much of the seemingly 'easy' concepts they are taught in schools actually took hundreds of years of collective thought to develop, with ideas and concepts gradually building within each generation of mathematicians and thinkers.While it is true that, the development of the mathematics has been subject to many extraneous factors (especially within Europe) such as the geopolitical landscape[8], higher priority tasks

[8] Many prominent mathematicians from the 18th and 19th century are French or German. Prior to WW2, the longest period of peace for both the French and German states were in - you guessed it - the 18th and 19th centuries particularly in the years prior to the French revolutionary wars (1763 - 1792) and the years following the Napoleonic wars (1815 - 1870). Many of the mathematicians we discuss in this text had the main portion of their careers in these periods such as Leonhard Euler (1707-1783), Adrien-Marie Legendre (1752 - 1833) and Carl Gauss (1777 - 1855).

upon educated civilians[9] or simply a plain lack of funding[10]. However, despite these extraneous factors the reason many problems that are seemingly trivial in textbooks have been unsolved is because they are just plain hard. It's very easy in hindsight to see how a problem may be solved but creating a novel solution is painstakingly difficult and requires hours of uninterrupted thought and reflection. Not to mention the prevalent case where one believes they have a correct solution which they later realise to be incorrect, requiring even further contemplation and reflection. At times fixing your incorrect solution can be harder than creating a new solution![11] Throughout history much of mathematical development has been a rampant case of trial and error. It's just that no one sees the trial and error so they believe that the end results materialise out of thin air.

Don't believe me?

---

[9] Many mathematicians had also studied or were practicing lawyers such as Leibniz, Cauchy and Descartes. Leibniz was actually involved in diplomatic efforts to counteract the growing power of Louis XIV. He sought to foster alliances between various European states to check French expansionism. His diplomatic missions and correspondences were aimed at creating a balance of power that would prevent any single nation, including France, from dominating Europe. It's unlikely with such pressing commitments, one would even consider trying to solve a mathematical problem[8].

[10] The post enlightenment period saw far greater societal value placed upon education and the broader sciences. Combined with the relative periods of peace post enlightenment, saw greater overall funding for mathematics[9].

[11] A famous case of this is in 1889 when the French mathematician, Henri Poncaire, submitted a manuscript for the three body problem which involves predicting the motions of three celestial bodies based on their mutual gravitational interactions. King Oscar II of Sweden and Norway offered a prize for anyone who could provide a solution to the problem and Poncaire's manuscript was initially awarded the prize when it first appeared to meet the requirements of the problem. However upon closer examination by Swedish mathematician, Lars Edvard Phragmén who was an editor of the journal Acta Mathematica that the manuscript was to be published in, Lars noted a major error in Poncaire's reasoning. Lars then notified Poncaire who acknowledged the area which then took several months to fix before final publication in 1890.

Allow me to introduce you to the archetypal mathematical examples of how long it really can take to arrive at the correct understanding keep in mind these are mathematicians that dedicated their *entire lives* to the field:

Andrew Wiles (1953 -) : A British mathematician who took seven years of work on the modularity conjecture which gave a conclusive proof of Fermat's last theorem.[12] A key note however is that Wiles worked on the problem in secrecy, with greater collaboration the solution may have been attained sooner - nonetheless *seven years* [10].

[12] Fermat's last theorem is a famous problem within number theory which states that for integers x,y and z the equation x^n + y^n = z^n has no positive integer solutions for n > 2. The problem was first formulated by Pierre de Fermat, a French mathematician in 1637.

Grigori Perelman (1966 -): A Russian mathematician who again took seven years to solve the Poncaire conjecture.[13] He established the groundwork for the proof of the conjecture during a two year visiting period at the University of California, Berkeley. Then virtually disappeared from the mathematical community for almost a decade in order to solve the conjecture[14].[11]

[13] The Poincare conjecture is as follows: Every three-dimensional topological manifold which is closed, connected, and has trivial fundamental group is homeomorphic to the three-dimensional sphere. The claim of the conjecture in plain english is that any object without a hole in it can be reshaped into a sphere.

[14] If you're wondering how he financed himself over the seven years he was living off of a combination of his mother's income and the money he had earned as a visiting researcher at the University of California, Berkeley.

Omar Khayyam (1048 - 1131): A Persian polymath who spent a significant portion of his life trying to find a general solution to the degree three polynomial[9]. He said in the later stages of his life that: "We have tried to express these roots by algebra but have failed. It may be, however, that men who come after us will succeed.[15]"

Now it's not important to the reader to understand what these mathematicians actually worked on (which spans a variety of

[15] And succeed they did roughly 400 years after Omar Khayyam's death by Italian mathematicians Scipione del Ferro (1465 - 1526), Nicocolo Tartaglia (1499 - 1557) and Gerolamo Cardano (1501 - 1576). Mathematicians in Italy in this time period were known to have mathematical duels where both mathematicians would give a set of problems to the other - the victor being the one who could solve all of their opponents problems. One of the reasons the general solution to the cubic evaded publication was because it held the secret to winning such duels!

different fields). It *is* important, however, to understand how long it took them and this is all while working on isolated problems within very specific fields. Try advancing the understanding of a whole field, upon many problems and subject matters and soon one can see how it truly does take years of collective effort to arrive at the understanding of mathematics we have today. Not to mention the level of correspondence (you would be surprised at how much even 17th century mathematicians discussed ideas with each other[16]) and communication that mathematicians undergo to prove their ideas. Nowadays mathematical papers normally have multiple authors with single author papers being highly uncommon[13].

We can observe then that the process of discovery, innovation and field advancement is a thoroughly time consuming process. The cases we cited above are actual extensions upon the field of mathematics and if we consider a standard school classroom, basic concepts can take months or even years to properly integrate into one's mind.

So if we really do acknowledge that it can take generations to arrive at the understanding we currently have within mathematics, is there anything we can learn from the experience of the generations of mathematicians that have come before us?

---

[16] Gottfried Leibniz (1646 - 1716) was a German philosopher and mathematician who spent most of the later portions of his life working on compiling a historical piece known as the 'History of the House of Brunswick'. Part of the reason this history was compiled was to settle land dispute claims that the house were having within Italy by claiming ancestry from particular regions. In compiling this work Leibniz was given a stipend and the opportunity to travel to different parts of Germany (then the Holy Roman Empire) and Italy to analyse archival documents related to the ancestry of the House of Hanover. As you might have guessed Leibniz spent most of the travels speaking to his contemporaries of his time including Baruch Spinoza, Christian Wolff and Antoine Arnauld. His work on the House of Habsburg was incomplete at his time of death - Leibniz cared more for collaboration than his assigned task![8].

Despite the hours of contemplation and thought, does there exist some kind of structure to this intellectual madness that we can extrapolate to support our own learning?

Though it may seem that all the advancements not just within mathematics but the fields of physics, chemistry and biology (all the sciences really) come randomly there *does* in fact exist a pattern to what we are observing. All mathematicians and scientific thinkers do undergo key stages in the development of their ideas, though these stages can be years long and the problems so intellectually demanding, that mathematicians are often not self aware enough to know what stage of development they are in - or even the time of their next meal[17] for that matter! Luckily through the study of their lives along with our own reasoning we can observe a visible structure - but it is all based around one key idea.

[17] Ramanujan was known for skipping meals[14].

Numerical Intuition

All advancements within the field of mathematics (and the broader sciences) can be reduced to one simple idea which we will define as numerical intuition:

Numerical Intuition: the inherent problem-solving ability one develops through repeated computation and attempts on problems.

Now upon reading this definition there are a few terms to unpack. The first is 'numerical' – upon reading the definition of *numerical intuition* one may notice that it actually has no immediate connection with numbers or numerals. The reason it is called 'numerical intuition' and not just 'intuition' or 'problem solving intuition' is because we will be analysing this property within the field of mathematics where a huge component of the understanding of the field of mathematics initially begins with numbers (numerals). Thus, it's logical to have a word that describes something which has some relation to numbers (numerals)[18].

The second reason is that the breadth of the corpus of current mathematical study really does change one's mind and perception

[18] Basically this book is about maths so I'm calling it 'numerical intuition'. The biologist down the street can call it 'biological intuition' though that sounds more like the feeling one has before they see a tiger in the wild as opposed to the fundamental idea underpinning mathematical learning.

to problems they may encounter[19]. When one has worked with certain mathematical objects for years (matrices, groups, metric spaces) they can develop a kind of intuition in problems that involve such objects - and since numbers are the most fundamental object within mathematics we will reference any intuition developed within the various fields of mathematics as numerical intuition.

The next is 'intuition'. Intuition in itself is a very unique characteristic which is recognised universally in virtually every culture. As it's understood, intuition is a feeling that one has upon the correctness of a decision or undertaking which can arise both from spontaneous feeling or from past experiences. Where it comes from is unknown and some scholars/spiritualists claim it comes from the workings of the subconscious mind. Either way many mathematicians acknowledge its existence as a driving measure for solving yet unsolved problems[20]. This makes total sense as one needs a target to direct their intellectual efforts toward especially if a potential solution can take years to arrive at.

Thus, when we put the two words 'numerical intuition' together based of the understanding we have of the words 'numerical' and

---

[19] This ability is not only isolated to mathematics. Any endeavour entails a degree of intuition built within its field - even playing a professional sport such as soccer. Even a simple act as crossing the ball into the box for a headed goal, requires implicit computations that the mind performs often without any conscious effort. Such an act also has the added dimension of predicting where the player can move to, the speed of the cross, loopiness etc and one can see elements of intuition at play.

[20] To the point where some mathematicians such as Henri Poncaire (1854 - 1912), renowned for his contributions to a number of mathematical disciplines, have even written essays about it - case in point is the essay 'Intuition and Logic in Mathematics' written by Henri Poncaire in 1905.

'intuition', what we arrive at is a phrase which means something along the lines of:

'numerical sixth sense'
'mathematical feeling of correctness'
'intuition relating to numerals'

Hopefully this explains to the reader why it's called specifically 'numerical intuition'.

Inherence

Upon looking at the definition again the reader may notice the phrase 'inherent problem solving ability' and wonder why the word 'inherent' is included?
The reader may wonder that the definition could potentially be written as below –

Numerical Intuition: the problem solving ability one develops through repeated computation and attempts on problems.

The reason the 'problem solving ability' is inherent is because it comes *automatically* through repeated computations/attempts on problems. The problem solving ability that arises through rigorous practice cannot be removed from performing repeated computations and is thus *inherent*. The nature of the human mind works such that the problem solving ability will necessarily come to a practitioner if they are repeatedly performing computations on any kind of mathematical problem. Now there exist various theories that attempt to explain how the human mind works in relation to learning - but everyone can agree on one fundamental principle which we will call the improvement principle:

Improvement Principle: the more you do something the better you get at it.

The improvement principle is a broad generalisation of neuroplasticity, which refers to the brain's ability to reorganize and adapt in response to experiences, learning, and environmental changes. This means that learning physically changes the structure and function of the brain by forming new neural connections and pathways. Another common analogy used is that the brain functions like (and actually is) a neural network which through the appropriate parameters, will necessarily improve its trained abilities (predictions, error minimisation etc).

Now the reason the inherence to numerical intuition is so important is that it lies beyond any innate or presumed mathematical ability one might or believe they have. From definition alone performing repeated computations and attempting problems one *will* necessarily improve their problem solving ability. As many would otherwise claim - it has nothing to do with any proclivity to performing mathematical calculations (which some children or school students may randomly possess), your upbringing, the school you attend or any societal factors. This is interesting because the way the mathematical education system is framed (especially nowadays with people showcasing their maths and coding skills on social media) people tend to believe that their mathematical capability is not developable. Well with the definition we have above the *inherence* to numerical intuition offers any practitioner a path to improvement within the field of mathematics.

Problem Solving Ability

Now we come to the actual skill which is being developed within numerical intuition which is problem solving ability. It's important for us to define what this actually means since problem solving ability can be quite ambiguous and is a term that is thrown around in various corporate and classroom settings.

In the context of this book, we will speak specifically about mathematical problem solving so when we say 'problem solving' it can be taken as mathematical problem solving. Either way we will still provide a definition for mathematical problem solving as it may not be clear to the reader what this actually is. Warning - the definition is self explanatory and trivial but it must be said to avoid any ambiguity in the contents of this text:

Mathematical problem solving: your ability to solve mathematical problems.

One aspect of the definition to clarify (though the definition is trivial) is that when we say 'mathematical problems' this is actually a very broad term. It can encompass any problem from number theory, statistics, topology to calculus and all the numerous subfields of mathematics.

Yet we still have the reference to 'mathematical problem solving' and not those of specific fields of mathematics such as 'topological problem solving', 'statistical problem solving'. This is because, despite the breadth of the field of mathematics, numerical intuition can be developed across all fields and not just on any specific field. Through repeated computations and attempts on problems in one area of mathematics, a practitioner improves their overall mathematical problem-solving ability which can be applicable beyond the area where the practice was attained. This is extremely important and a very interesting property of numerical intuition which we will elaborate on in the section *Properties*. What we are essentially saying is that development in one area of mathematics can support problem solving in other areas as well hence one term suffices to describe the development of problem solving ability as related to mathematics - *mathematical problem solving*.

Repeated Computation/Attempts

In school or any mathematical institution when one adds two numbers, solves a linear equation or finds the derivative of a basic function they often say they are 'doing' maths. This however is a misnomer. We can also tell a computer to do the exact same procedure when adding two numbers, solving a linear equation or finding derivatives but you wouldn't say that a computer is 'doing maths' right?

You may say instead that the computer is *computing* or rather *performing computations.* Most of the basic mathematical procedures we are taught are *computations*, there are pre-defined precise steps that lead to a definite answer. Within schools, universities, or any time you're studying maths as it is conventionally taught, you are simply repeatedly performing computations. Though you may think you are getting better at 'maths' the skill being improved in all of this is simply your computing ability. However once your computing ability reaches a certain point, one begins to develop a 'feel' or *intuition* for what an answer could be or look like (at least by order of magnitude range). It is *this* ability which is numerical intuition which is only obtained from performing all the repeated *computations* and going through all these intermediary learning steps.

To highlight the contrast between 'maths' and 'computation' suppose I gave an equation to student A and another to student B.

Student A takes 5 seconds to solve it and Student B takes 1 minute. The teacher/fellow students will then go on to say that Student A is better at maths! What should instead be said is that student A has greater computing ability which is easily developable. Now going one step further and suppose I ask another question to Student A and Student B where the answer (just suppose) is 10. Now again, suppose we ask student A & B what they think the answer should be. After inspection student A says -5 and student B says 8. In this scenario we can say that student B has greater numerical intuition, being able to *feel* the answer out without necessarily having to compute it. The example highlighted here is just one example of numerical intuition supporting computation, as we will see in later chapters it also supports the overall understanding of a concept and innovative extensions.

A similar outcome arises from *attempting* problems; even failed attempts contribute to one's numerical intuition. At a fundamental level even with unsuccessful attempts one can eliminate at least one line of reasoning or an egregious method which sharpens your intuition and feeling for the correct solution on later problems. Even in more abstract areas of mathematics, where one does not perform arithmetical computations, attempting problems still functions like a computation but in the more abstract plane[21]. Thus,

---

[21] This is evidenced in some sense by proof checking algorithms. Proof checking algorithms are essential in various fields such as computer science, mathematics, and formal verification to ensure that given proofs are correct. These algorithms range from simple syntactic checks to sophisticated formal systems that verify the logical consistency of complex proofs. Here's an overview of key concepts, techniques, and types of proof checking algorithms:

not only computing but attempting (trial and failure) also functions to develop one's numerical intuition.

Summary

In this section of *Frame* we have covered some of the core fundamentals behind numerical intuition. We observed that there is a structure to the method of discovery and intellectual improvement within the field of mathematics leading ultimately to new insights and discoveries. We unpacked numerical intuition, its definition and the various components of its definition which were important to clarify – *inherence, problem solving ability, repeated and computation/attempts.* In the following sections we will now begin to analyse how numerical intuition is developed on a deeper level and some of the key factors that contribute to its development (beyond those in the definition). The key idea of the next section is how one can actually perform the required *computations* in new fields to improve their ability even when they haven't been exposed to the field before. This lies at the core of the difficulty of mathematics (think back to how difficult fractions, linear equations and derivatives probably were at first) and we will uncover what truly allows progress in an unknown learning domain. The answer is not what it seems.

Part 3: Sufficiency

John Conway (1937 - 2020): "Becoming sufficiently familiar with something is a substitute for understanding it."

Introduction to Familiarity

Within the coming sections we will analyse the contributing factors in the development of numerical intuition. To get a basic idea of what these factors are, we will refer again to the definition:

Numerical Intuition: the inherent problem-solving ability one develops through repeated computation and attempts on problems.

Now if we look at the definition, we will see that only one component of it is actually within the control of a practitioner; which is the 'repeated computations or attempts on problems' highlighted in bold. As this is the only real factor in control of a practitioner, anything related to the development of numerical intuition would by consequence be related to *performing repeated computations or attempts on problems.* Now the first step to being able to perform computations is being able to *recognise* the objects and their related rules for computations and this is present even at the earliest stages of mathematical learning such as with arithmetic addition.

By the word '*recognise*' in this context, I mean that a practitioner should be aware of what certain symbols stand for and represent for example sin, cos, $\frac{d}{dx}$, +, -, > and so on; in order to identify which computations to perform. It is evident that a practitioner must clearly *recognise* the object with which they are to compute, having defined recognition this way (how can a practitioner compute an object they cannot even recognise[22]?). While seemingly trivial, this leads us to one of the most important insights pertaining to mathematical learning. which we will give the name: *Difficulty by Nascency.* The premise of this insight is that the stage of *recognition* is of great difficulty for practitioners because of the nascency of the object they are to compute with. In simpler terms, all I am stating is the following:

Difficulty by Nascency: Mathematics is really only hard while it's new.

To see cases of *Difficulty by Nascency*; I encourage the reader to ask themselves, how many practitioners struggle when they begin to see the alphanumeric characters[23] within mathematics or how many practitioners struggle when they see the unusual notation of calculus (derivatives and integrals)[24]?

At a fundamental level, the actual performance of computations using these objects (alphanumeric characters and the notation of calculus) boils down to nothing more than following instructions. Furthermore at the primary and high school levels, virtually all

---

[22] Try teaching trigonometry before showing someone what a triangle is.

[23] Here I mean x's, y's and the numerous Greek alphabet letters used to represent values in mathematics.

[24] Even the usage of symbols to represent mathematical quantities is mostly credited to have happened *after* the Ancient Greeks. An entire civilisation existed before the introduction of the representation of quantities in this manner, so let it be known to the reader that virtually every concept in mathematics (even notation) is non-obvious before its creation.

problems/computations in these areas can be performed by a computer without even using approximations[25] since the solutions can be obtained *exactly*. This means that there exists a series of definitive steps a practitioner can essentially replicate, just as the computer does to solve problems at such a level. Now when a computer performs these pre-programmed steps to obtain a solution, does the computer actually know what it is doing? Does the computer know what a number is, or what a derivative represents? Though it may seem like it does, the computer is not conscious; these behaviours are already encoded into the machine so that it behaves in a specific way to input information which is something that a practitioner can also do.

Yet these problems involving alphanumeric symbols and calculus, for a human, have the perception of difficulty hovering over them and are - at times - purposefully avoided from the day they are taught. What practitioners at large do not realise is that, the *real* most difficult aspect of these commonly perceived 'difficult' problems (algebra, calculus, matrices etc) is simply the fact that they have not seen anything like it before i.e these problems are *nascent*. If we come back again to the definition of numerical intuition:

Numerical Intuition: the inherent problem-solving ability one develops through <u>repeated computation and attempts on problems.</u>

[25] If you're wondering why approximations are even a thing within the real world when we should have exact answers for all problems, some problems within mathematics can be so difficult to solve exactly that approximations are the best we can get. A quick study of integration and derivatives or even the irrationality of pi will show the intrinsic limitations on precision.

The only part of the definition that is within the control of a practitioner is the performance of repeated computations/attempts on problems. If a practitioner is completely new to a mathematical object, then they haven't performed any computations on problems related to the object. Thus the practitioner will have no inherent problem solving capabilities and hence problems involving the object are now apparently difficult. Instead of saying *difficult,* a more suitable term would be *unfamiliar.*

No one in the history of mathematics has understood an idea in its deepest sense without performing some kind of computation on the objects they are trying to understand. It's only within the current day and age that practitioners aim to bypass the stage of computation and believe they can obtain understanding without performing any rigorous practice[26].

So what we require is a way that we can actually perform repeated computations in new *unfamiliar* areas of mathematics, so as to begin developing numerical intuition in these areas. This is where the central idea of this chapter: *familiarity,* comes in; which we have touched upon above but we will define formally as follows:

Familiarity: the level of continued exposure one has to a concept.

[26] Early mathematicians including Newton, Leibniz and even Gauss performed all their computations by *hand.* There is even a famous story of Gauss helping his father out with some family business accounting when he was three years old. Talk about already developing numerical intuition.

This definition is the absolute cornerstone of mathematical learning and offers us a pathway to progression through unfamiliar learning domains. In mathematical fields that are very unfamiliar and highly abstract, it is normal for a practitioner to feel a degree of bewilderment upon first exposure to the concepts within the field (even if they are aware of the *Familiarity* definition above) and it must be said that this is completely *normal* to learning.

Normality of cluelessness

Through various stages of mathematical learning one can often find themselves absolutely clueless when it comes to a new subject matter, mostly due to *nascency* as we have discussed in the previous section. This is especially prevalent when new kinds of notation (think fractions, algebra and calculus) and subjects (linear equations, quadratics, vectors etc) are introduced to a practitioner. What's important to understand here is that it is actually *normal*[27] to be clueless when learning new mathematics. From primary school to university, almost every new concept within mathematics, I believe, actually should leave a practitioner clueless upon first encounter[28]. If a practitioner finds themselves in such a situation of complete perplexity then an important quote to remember is one from John Conway[29] which prefaced this chapter:

"Becoming sufficiently familiar with something is a substitute for understanding it"

This quote embodies the essence of this section of the book, but we will rephrase it slightly so it better reflects real life mathematical learning. Let us also give it a name, *the familiarity principle*:

---

[27] Albert Einstein: '"Do not worry about your difficulties in mathematics. I can assure you mine are still greater."

[28] Though not exactly strictly mathematical, Niels Bohr sums it up well - "Anyone who is not shocked by quantum theory [mathematics] has not understood it."

[29] John Conway (1937-2020) was a British mathematician renowned for his work in various areas of mathematics, particularly in the fields of group theory, knot theory, number theory, and combinatorial game theory. He is perhaps most famous for inventing the cellular automaton called the Game of Life, which is a zero-player game where an initial configuration of cells evolves over time according to specific rules. Despite its simplicity, the Game of Life has fascinated mathematicians, computer scientists, and enthusiasts for decades due to its complex, emergent behavior.

The Familiarity Principle: 'Becoming familiar with something is the first step for understanding it'

In primary and secondary schooling, practitioners will generally be taught concepts that are relatable to the real world but are not very abstract in nature, such as trigonometry or statistics. Thus, practitioners will usually be able to recognise (or visualise) some aspect of problems they are presented with and, as a consequence, will usually have some degree of familiarity with the problem. For example, if a teacher was to show even a year 8 or 9 student a derivative such as they would at least be able to *recognise* the equation in the brackets $x^2 + 7x + 1$. For the derivative shown a practitioner still has a level of exposure to at least *something* shown and this serves as the first step in order to actually understand and compute with the derivative. Imagine now that a practitioner does not know even such basic details, or is unable to recognise the notation in an equation given and *this* is the kind of situation any university mathematical student would find themselves in at some point. However, as virtually every university student does, they then apply the Familiarity Principle to learn an aspect of the problem, look/study the related concepts day after day, learn what the next component of the problem is and so on till they start to begin understanding the underlying concept.

Additionally the familiarity principle is not strictly a mathematical principle, it is also relevant to virtually every other field. Take the field of computer science for example. I can guarantee personally that absolutely no software developer understands how an electron's state encodes information within their programs, yet your everyday Silicon Valley developer will claim that they understand, in its entirety, what their computer program is actually doing. Let's be clear, the programmer only really knows what is happening at the highest level which could include the computer's memory, CPU and RAM at best. Essentially, the understanding is limited to only the functional requirement of the program or whatever is relevant to their work.

If the program was creating a website for example, then the programmer would know only basic details such as how changing the code would affect what is displayed on the website - but that's it really - only to the level of the functional or business requirement is the program understood. Beneath this level of understanding most developers would have no idea how their program continues down the chains of abstraction in computing. In reality, when the programmer runs the code it is then formatted (compiled) in a way the computer can actually understand. Then different parts of the computer (CPU, RAM and memory) actually work to physically compute and execute a response to the program with the given input. To execute this response, electrons are moving through transistors that encode the information initially carried by the program.

We can observe in this example, that there exist many layers of abstraction within running a basic computer program and a programmer likely will not know - to the level of the electron - how their program is running, but they are *familiar* enough with the high level components of their program to fulfill the functional requirements they may be tasked with. Better yet, they can still *improve* their programming skills without actually knowing to the greatest detail how a computer program actually runs. This kind of principle, of working without a complete understanding, is one we will investigate in the coming chapters.

However it must be stated that even with an initial incomplete understanding, through targeted questioning and research a programmer actually *can* begin to understand how their program is run. By having a degree of familiarity with the program, the programmer can then ask with such questions as:

Why do we have to compile a program?
Why does this use machine code and not use instructions in plain English?
How does an electron encode information?

By answering such questions, having first obtained contextual familiarity behind the questions, the programmer can start to understand how their program is run in its entirety. So we can observe here how becoming familiar with a concept leads us to targeted questioning/research which allows us to then understand the concept, thus the familiarity functions as the first step toward understanding. We will analyse the relationship between familiarity and understanding more in the second book of the series *Numerical Intuition: Understanding*. For this text however, simply knowing that cluelessness is the convention and familiarity allows us to bridge this cluelessness is all we need.

Progression through Familiarity Alone

It's great that the familiarity principle gives us a way to begin understanding any new concept. However as we will see, for *progression* within a mathematical field a practitioner does not even require understanding - simply having enough *familiarity with a concept* is *sufficient.* In essence, much like how a programmer that is familiar enough with their software (but does not fully understand every component of it) can still improve their programming skills, a practitioner can progress in a mathematical field without having a full understanding of the field's concepts. It is important here that we clarify what exactly is meant by the term 'progression' or the phrase 'progression within a field' to avoid any ambiguity:

Progression: To improve numerical intuition within a field.

Note: In developing this ability a practitioner is 'progressing'.

An important point to clarify following the introduction of this definition is that a practitioner's progression within a field does not necessarily mean that they have to be performing anything more advanced. If we think about making 'progress' on a construction project (like a skyscraper) or university assignment it is implied that a practitioner is working towards completion. Similarly, if a practitioner learns all the different kinds of throws, locks and holds in a grappling martial art we can say that they are becoming more 'complete' grapplers. Within becoming more 'complete', a practitioner is obtaining coverage over the *scope* of their field and can thus be 'progressing'. We can refer to this kind of progression as *progression by scope* which is when one progresses by covering the *scope* or breadth of their respective fields.

This helps us to understand how progression within a field such as music is different to progression within mathematics. In one case a practitioner may be learning to play a greater variety of different pieces of music to be 'progressing' whereas in mathematics one is generally moving to more advanced stages of application in a specific area to be 'progressing'. This second kind of progression we can refer to as *progression by advancement* which we will cover in the coming sections.

In order to understand progression using familiarity, let us look at a non-mathematical thought experiment. Imagine, for a moment, that a ten year old kid was playing basketball but was not familiar with the rules. Perhaps they only know that the game involves a ball and its placement in an elevated hoop. Now imagine such a ten year old playing in a basketball game (which in mathematical terms could be an examination). It would be fair to say that the ten year old would probably be terrible within the game right? It is likely they would foul repeatedly, double dribble, travel and constantly keep, *well* breaking the rules. However, if we then teach (i.e familiarise) the kid the rules, he/she will then adapt and become a better player. In accordance with the rules they can practice and build their specific skillset (dribbling, shooting, running etc). A player does not even need to understand the reason behind the creation of the rules in the way they are, yet they can still work within the bounds of the rules and improve as basketball players. To understand why the rules are constructed the way they are (i.e having understanding) is not necessary for the players to improve. In this particular example it is almost nonsensical for a player to understand why the rules are constructed in the way they are, since it has no relevance to their actual performance within the sport. Yet prior to an exam (mathematical equivalent of a basketball game), practitioners study all night to try to understand every concept they may encounter not knowing they are functioning like a group of basketball players trying to understand why double dribbling is not allowed. Thus we can observe through kid-basketball examples how familiarity alone is sufficient for improvement and progression (an understanding of the rules of the game is completely irrelevant to actual performance just like how understanding a mathematical concept is, at times, completely irrelevant to solving a problem involving it).

This idea of attaining familiarity **first** to progress within a certain domain of learning can also be extended to other domains such as reading (one must be familiar with the letters), chemistry (being familiar with different substances and rules), piano (familiarity with chords and different pieces of music) and the list goes on. In virtually any new concept or learning environment one by nature will be unfamiliar with the subject matter. The important thing in such an environment is continual exposure. An equation or other mathematical object can appear to be difficult (even daunting initially), but look at it 10 times, 100 hundred times, 1000 times, even 10 000 times and then see how it gradually becomes easier to progress to other more difficult objects.

Progression by Scope (Mastery)

Now to further observe how familiarity alone is sufficient for improvement and progression let us look first at the case of progression by scope, which we can define as follows:

Progression by Scope: to develop greater (numerical) intuition with a concept through the scope of problems involving it.

A good way to think about this is that when people often talk about the Malcolm Gladwell 10,000 hours rule (mastery in anything takes 10,000 hours of dedicated practice), they are indeed *progressing by scope.* Now the reason there are parentheses around the word (numerical) is that this principle extends far beyond mathematics.

The large majority of the discourse on mastery centres around fields that are very specific and definite regarding the practice of their related skill sets. What we see in the current day is that much of this discourse is centered around, I would argue only the fields sports, chess and music[30] which are all relatively simple fields to

---

[30] I really do mean *only* sports, chess and music. Here is a list of savants and 'geniuses' that are commonly considered when thinking about thousands of hours of practice:

Lebron James - Basketball
Michael Jordan - Basketball
Lionel Messi - Soccer
Muhammad Ali - Boxing
Tom Brady - American Football
Bobby Fischer - Chess
Magnus Carlsen - Chess
Garry Kasparov - Chess
Ludwig Van Beethoven - Music
Wolfgang Amadeus Mozart - Music
Jimi Hendrix - Music

practice skills within. For instance, in sports one can repeatedly perform common motions such as throwing, kicking a ball, catching etc to improve their skills (other forms of sports are harder to practice individually, such as rugby, but many sports reduce to this kind of one-skill centred practice). Within chess, a practitioner can observe move configurations and play thousands of games, after having learned the basic rules of chess. Within classical music, one practices by learning to play different compositions of music with an instrument and works toward perfecting each individual piece. In the case of sports, music and chess one can see how 'practice' is very well defined, without there being many abstract elements involved. Contrast these fields to trying to practice writing (literature), artistic endeavours (painting, sculpting, drawing etc) and even music *production* (completely different in nature to practicing the instruments *involved* in the production of music) which are all highly abstract; and a practitioner can see how the nature of practice varies depending on the field. If we treat the most basic parts of these fields as 'objects' (chess moves, three pointers, musical notes) similar to how we view mathematical objects such as numbers, shapes etc, then we can say that a practitioner is improving in these fields by practising the possible configurations i.e the scope, in which these objects are *involved.*

Freddy Mercury - Music
Luciano Pavarotti - Music

The list can go on indefinitely.

To elaborate on what I mean here and of improvement through scope in general, let us look more deeply at the example, again, of sport. In this instance, let us examine the field of American Football (NFL) and the 'object' of throwing the ball for a quarterback. When training to become a quarterback, a practitioner works predominantly to improve their ability to throw a football. This includes throwing the ball accurately, over long distances, at different angles, speeds etc which are all part of the 'scope' of ways in which a football can be thrown. The ability to cover this scope is what constitutes improvement for the quarterback, as they can then cover all the potential scenarios that may appear in a real game. The same can be said of basketball and the basic 'object' (foundational element) of shooting the ball into the net, where a practitioner spends practically all of their time shooting the ball from different parts of the court[31] - covering the 'scope' of all the different situations they may shoot in within a game.

In each of these fields a practitioner progresses by practice in a more permutational sense - by performing computations/attempts on problems with all the potential situations their specific object can appear. Hence the practitioner develops their numerical intuition in these fields through the *scope* of potential problems with any one object and improves their inherent problem solving ability for any problem they may encounter which involves the object.

[31] I think the most memorable and relatable case to what we are discussing here is when Kobe Bryant once remarked that he would take 1000 shots every practice. In the context of this book, this is a strong case of progression by scope - I guarantee those 1000 shots included all the possible ways in which he would have to potentially take shots within a game.

Progression by Advancement (Progressive Links)

So we know that there exists progression that occurs by covering the scope of problems with an object (in a field). However there is one important note within this case - which is that, while there is definitely *progression* there is no *advancement* in *progression by scope*. One that is progressing in this manner is not necessarily progressing to a more advanced area of a field, but is becoming more proficient in concepts they are already familiar with. This, is still very significant, since proficient skill relating to an object or ability is a remarkable achievement in and of itself[32].

So the question then remains of how one is to progress in a field where something *new* or more advanced is unfamiliar. Now are there any mechanisms that we can extrapolate from familiarity, thus far, to derive a process for *advancement?*

Let us first define the manner of progression which we desire and then see if can derive a mechanism through familiarity:

Progression by advancement: to build familiarity with concepts in a field (that you do not already know).

[32] Case in point here is Stephen Curry of the Golden State Warriors whose proficiency in three point shooting completely revolutionised the game of basketball. His ability to perform three point shots across a number of different scenarios (including very long range three point shots) is indeed a remarkable ability and we see through this case how progression by scope, though not technically containing any real 'new' element (it can be argued that shooting three pointers the way he does *is* actually 'new' but the point of emphasis here is that he did not *invent or rather popularise* a new shot like Kareem's Skyhook or Nowitzki's One-Legged Fadeaway, but rather perfected an already existing shot - the three point shot) , is still incredibly relevant to many fields, sporting being one of them.

As we shall see this can be accomplished through numerical intuition.

Within maths especially and the broader sciences, a more 'difficult' or 'harder' area of study will usually extend on some previous understanding. Take for example the process of photosynthesis. Most people know that this is the process of converting carbon dioxide and water into glucose and oxygen and most classes will initially teach the basic equation below:

Photosynthesis Equation: $6CO_2 + 6H_2O \rightarrow C_6H_{12}O_6 + 6O_2$

But what about all the other related processes? Water Splitting, Electron Transport Chains and ATP formation (not to mention various others)[33] are also components of photosynthesis which aren't usually taught but all have a direct relation to the base equation:

Water Splitting: the process in photosynthesis where water molecules (H_2O) are broken down into oxygen (O_2), protons (H^+), and electrons.

Electron Transport Chains: A series of protein complexes and electron carriers embedded in a membrane (mitochondrial inner

[33] Listed above are the processes that are dependent on light (photons) in some capacity. There are also processes involved in photosynthesis which are performed without the presence of photons: carbon fixation, the reduction phase and regeneration of RuBP.

membrane in respiration, or thylakoid membrane in photosynthesis) that transfer electrons step by step.

ATP Formation: the process where energy stored in a proton gradient across a membrane is used by the enzyme ATP synthase to convert ADP and inorganic phosphate (Pi) into ATP.

How is it then that one can draw a connection to these more advanced chemical reactions using familiarity with the base photosynthesis equation?

In general, any advanced subject matter will be composed of smaller components which the practitioner will already have some familiarity with. To build on the photosynthesis example above, a related process is water splitting which is where the water molecule in photosynthesis turns into oxygen, protons and electrons. Now a practitioner may not exactly know how water splitting works, but simply knowing that oxygen, protons and electrons are produced from it will at the very least aid a practitioner to be able to *identify* this process. Now the practitioner knows that the basic photosynthesis equation has an H20 molecule but also 'sunlight' or light as a catalyst and can possibly deduce that the 'splitting' occurs from the interaction between the water molecule and sunlight. So what happened here is that a more complicated procedure (water splitting) was broken down into components which we are *familiar* with (H20 and sunlight) and thus realising the link between the H20, sunlight we were able to understand what water splitting does through familiarity with its component parts.

The actual mechanism behind the splitting of the water molecule again is not understood within this frame, but we are familiar enough to begin computing/working with problems that may involve water splitting.

Thus in general, we can apply familiarity with simple subjects and be able to progress to advanced subjects. This is where one has to realise the link between the simple and advanced area. In fact we will give this link a formal definition:

Progressive Link: the additional information required to work with concepts you are unfamiliar with.

Note: 'work with' in a mathematical context would be performing computations or solving problems involving a concept.

Often the simple subjects form a *layer of understanding* that a practitioner would already be familiar with, all that is required to then begin computations or to work with advanced subjects is to learn the link to the advanced/difficult subject from the layer of understanding the practitioner possesses.

Sufficient familiarity

So we know then that one can progress within a mathematical field through familiarity alone. This occurs through the cases of *progression by scope* and *progression by advancement* as discussed in the previous section. So with these two ways of progression established - one naturally resulting question is the following:

'How much familiarity is required with a component of a field to progress then to another related component which is more advanced?'

Now this may sound like a complicated question but the following questions listed below are examples of the question above pertaining to particular fields and components:

'How familiar does one need to be with a triangle (component) before one can progress to sin, cos and tan usage (related components)?'

'How familiar does one need to be with Netwonian gravity (component) before they can progress to General Relativity (related component)?'

'How familiar does one need to be with differential calculus (component) before progressing to differential equations (related component)?'

'How familiar does one need to be with fields (component) before progressing to Galois field theory (related component)?'

And to give examples that are not mathematical, to really emphasise the core meaning of these questions:

'How familiar does one need to be with a cell's structure (component) before progressing to mitosis/meiosis (related component)?'

'How familiar does one need to be with DNA (component) before progressing to protein synthesis (related component)?'

'How familiar does one need to be with amino acids (component) before progressing to enzyme activity (related component)?'

And to give examples that have no relation to the sciences to *really* emphasise the base question:

'How familiar does one need to be with financial regulations (component) to become an accountant (related occupation)?'

'How familiar does one need to be with drainage systems (component) to become a plumber (related occupation)?'

'How familiar does one need to be with human anatomy (component) to become a physiotherapist (related occupation)?'

And you get the gist.

In this excerpt we will establish the level of familiarity required with any component of a field for advancement (this idea extends also to non mathematical fields), so we will actually formalise the level of familiarity required for progression and thus finding the answer to mathematical question such as:

'How familiar does one need to be with linear equations (component) before they can progress to quadratic equations (related component)?'

As well as various non-mathematical and even non-scientific counterparts of the same question such as:

'How familiar does one need to be with a car engine (component) to become a mechanic (related field)?'

As the underlying principles within both questions are actually the same.

What we know thus far in relation to familiarity is that familiarity is a measure of the level of continual exposure to a concept, and we also know from our discourse on numerical intuition, that numerical intuition is developed through *repeated computation/attempts* on problems. What this means is that to continue developing numerical intuition (which allows us to continually progress within a field) one must be able to perform repeated computations and should have enough familiarity with new objects to be able to perform computations on them. To give an example - trigonometric functions may look quite daunting when first encountered, but fundamentally a practitioner can refer to trigonometry tables to find the values of sin(45), cos(90) etc. One may have absolutely no idea of what they are actually calculating but they should be *familiar* enough with the notation and requirements of the problem to know how the solution is found/calculated. With this base understanding one is then *sufficiently* familiar to be able to progress to more advanced stages of learning, simply by being able to perform simple computations:

Sufficient Familiarity: the minimum level of familiarity required to perform computations and attempts on problems.

Thus what we are saying is that in order to develop numerical intuition, a practitioner only requires familiarity sufficient enough to be able to perform repeated computations (or at least attempt a problem). Through the repeated computations a practitioner develops their numerical intuition, and when they progress to more advanced components of a field they will, at least, be able to perform computations on the simpler components for which they have intuition.

To give an example from a well known area of mathematics let's look at fractions. Suppose that a class of practitioners had learnt addition and multiplication (with numbers) and then you give them the following fractions problem:

$$\frac{3}{4}+\frac{1}{3}$$

In order to perform this problem the practitioners must know that to first the lowest common multiple of 3 and 4 which is 12. This first step is the link between the numbers (simpler component) and the fraction (advanced component). Once this step is realised the problem reduces to one the practitioner is *already familiar* with. Namely, at this stage the practitioner will then multiply the top numbers to give equivalent values of each fraction i.e the practitioner will perform simple multiplication which is an operation they are already familiar with:

$$\frac{3 \times 3}{12}+\frac{1 \times 4}{12}$$

And then the practitioner will perform an addition operation with the numerators which is also an operation they are already familiar with:

$$\frac{9 \times 4}{12}=\frac{13}{12}$$

So really how much did a practitioner actually need to know about fractions to perform the above operations? Not once did we give an analogy to understand the nature of fractions (like pizza) or try to explain what a fraction actually represents to perform this operation. What happened is that we knew the *progressive link* between number computation and computation with fractions (make the fractions have common denominators) and we were *sufficiently familiar* with number computation to obtain the final answer and this *alone* allowed us to solve the problem of $\frac{3}{4} + \frac{1}{3}$.

Now the example of numbers and fractions is perhaps the first a practitioner would encounter but the same idea extends to virtually all objects within mathematics. So coming back to the question:

'How familiar does one need to be with linear equations (component) before they can progress to quadratic equations (related component)?'

Well the answer is that they must know the progressive link between quadratic equations and linear equations (factorisation) and then be sufficiently familiar with solving linear equations.

And for the non-mathematical example:

'How familiar does one need to be with a car engine (component) to become a mechanic (related field)?'

Familiar enough to know the link between a customer's query and the engine's potential problem, and sufficiently familiar with an engine to make the required adjustment.

Another key point to reiterate is this idea of sufficient familiarity does not require any intelligence or IQ. I cannot understate the importance of the previous sentence. To emphasise it even further; we can tell a brain dead computer with no consciousness how to add two numbers, perform integrals and matrix operations which are all done by following predefined instructions in a computer.

Similarly with humans, if one is simply familiar enough to recognise how computations can be performed (even if they have absolutely no idea what they are doing) this is sufficient in developing numerical intuition. To summarise, if a human can identify the relevant mathematical object (two numbers, a fraction, a decimal) - any mathematical object on which we are performing the computation, then computations on more advanced and difficult fields can then be understood once their link to the simpler computations is established (*progressive link).*

Thus what we can observe is that complete understanding is not required to be able to progress within a mathematical domain. All we need is to be familiar enough (*sufficiently familiar*) to be able to recognise how computations are performed on the relevant object, and with this we can move to more complicated fields of study through *progressive links*, where we can apply the same principle again of performing computations through *sufficient familiarity*.

Sufficient Familiarity and Numerical Intuition

So we have covered, thus far, the concept of familiarity and its related concept of sufficient familiarity. Let us now take a moment to synthesise the relation between these two concepts.

The first idea was *numerical intuition* within which we require the performance of repeated computations to develop inherent problem solving ability. This is trivial with very basic mathematical concepts within early schooling, however when concepts become more difficult there can be greater obstacles to understanding. So what we then require is a way to continue computing and developing our intuition without having full understanding of a concept.

Then comes *familiarity* is the level of continual exposure we have had to a learning domain or a concept within a field. Familiarity allows us to recognise symbols and different kinds of notations along with the rules that apply to performing computations. Sufficient familiarity, more specifically, is the minimum level of familiarity we require to perform a computation on the symbols and objects we are able to recognise. Sufficient familiarity coupled with the related progressive link allows a practitioner to continue performing computations on advanced fields by learning the link between the basic and advanced components. This can be surmised in the equation below which we will call the *Progression Equation*:

Progression = Sufficient Familiarity + Progressive Link

All this information will, at this point, likely appear as empty words to the reader. The examples I have given thus far to illustrate the points proposed are mostly from the basic sciences and mathematics. Beyond this point we will instead look at the lives of different key scientific figures who, within specific portions of their careers, have shown instances of the application of the ideas we have discussed thus far.

Sufficient Familiarity: Charles Darwin [15]

Though numerical intuition is the topic of this book, certain *scientific* figures give prime examples of the applications of the same principles underlying numerical intuition. The sciences offer more well known examples of figures who have had an incomplete understanding of their theorems at the time of proposal and none better than the case of Charles Darwin, the 19th century naturalist who revolutionised our understanding of human evolution.

Charles Darwin (1809-1882): The 19th-century naturalist who formulated the theory of evolution by natural selection, fundamentally transforming our understanding of the diversity of species we observe on Earth.

Now the reader might be wondering how non-mathematicians showcase the principles of numerical intuition. A key note here is that, while numerical intuition relates more closely to mathematics there is still one aspect of the definition which is not explicitly mathematical:

Numerical Intuition: the problem solving ability one develops through repeated computation and attempts on problems.

By focusing on the '*attempts*' component of the definition we can then see how Darwin applied the foundational concepts of familiarity. The main point to note about Charles Darwin is that there was no understanding of genetics, DNA or mutations during his lifetime[34] and yet, Darwin was still able to explain many of the core principles of evolution. In our terms, he was able to *progress* to an accurate explanation without having a full understanding of the mechanisms within his own theorem!

Now to give credence to any skeptics, it is plausible to make the argument that Darwin was presenting more of an idea than a rigorous theorem. Nonetheless Darwin still did not understand the core mechanisms that made his discoveries possible and observable within nature. Mutations which are at the core of

[34] Oswald Avery, along with colleagues Colin MacLeod and Maclyn McCarty, provided evidence that DNA is the substance that carries genetic information. This discovery was a pivotal moment in molecular genetics and was discovered in 1944, over 60 years after Darwin's death.

hereditary trait transfer, underlie evolution as we know it today and this central idea was unknown to Darwin.

Now, not to discredit Darwin's achievements; his experiences and scientific works cannot be surmised within a few short pages. Instead what we are going to emphasize here is how Darwin applied the principles of sufficient familiarity to natural selection, making attempts at explaining his observations despite *having an incomplete understanding of* the theory of natural selection at the time.

Short Biographical Account of Charles Darwin

Charles Robert Darwin was born on February 12, 1809, in Shrewsbury, Shropshire, England, into a prominent and well-to-do family. His father, Robert Darwin, was a successful physician, and his grandfather, Erasmus Darwin, was a renowned physician, naturalist, and poet. From an early age, Darwin showed a keen interest in the natural world, often collecting specimens and observing plants and animals in his surroundings. Despite his initial interest in becoming a doctor like his father, his passion for natural history eventually led him to pursue a different path.

In 1825, Darwin enrolled at the University of Edinburgh to study medicine, but he found the lectures uninspiring and the surgical demonstrations disturbing. After two years, he left Edinburgh and transferred to Christ's College, Cambridge, with the intention of becoming a clergyman. At Cambridge, Darwin's interest in natural history was nurtured by his professors, and he became particularly influenced by the ideas of the geologist Adam Sedgwick (one of the founders of modern geology) and the botanist John Stevens Henslow (who worked in plant classification). In 1831, Darwin received an invitation to join a scientific expedition aboard the HMS Beagle as a naturalist. The voyage, which lasted nearly five years, took Darwin to various locations around the world, including South America, the Galápagos Islands, Australia, and South Africa. During this time, Darwin collected numerous specimens and made extensive observations of the geology, flora, and fauna of the regions he visited. It was during the voyage of the Beagle that Darwin began to formulate his theory of evolution by natural selection. He observed variations among species and noticed how environmental factors influenced the survival and reproduction of organisms. These observations laid the groundwork for his revolutionary ideas about the origin of species.

After returning to England in 1836, Darwin spent many years analyzing his notes and specimens, as well as corresponding with other scientists. He also conducted experiments and collected further evidence to support his theory of evolution by natural selection. In 1859, Darwin published his seminal work, "On the Origin of Species by Means of Natural Selection," in which he presented his theory of evolution and provided evidence from various fields, including paleontology, embryology, and biogeography. The book caused a sensation and sparked intense debate among scientists and the general public. Despite facing criticism and controversy, Darwin's theory gradually gained acceptance and became the foundation of modern evolutionary biology. Throughout his life, he continued to research and publish works on topics such as botany, geology, and the expression of emotions in humans and animals.

Charles Darwin died on April 19, 1882, at the age of 73, leaving behind a legacy as one of the most influential scientists in history. His ideas revolutionized our understanding of the natural world and continue to shape scientific inquiry to this day.

Darwin and Sufficient Familiarity

Now let's analyse how Darwin was able to come up with natural selection through our understanding of sufficient familiarity. We first know that Darwin was travelling through the Galapagos Islands on the HMS Beagle for five years. Let's emphasise this point one more time in greater detail. Darwin was aboard a ship studying and observing the environment around him for *five years.* And what exactly was around him? Here is a list of the geographical areas Darwin covered as part of his five year journey on the HMS Beagle:

England
Brazil
Argentina
Uruguay
Falkland Islands
Chile
Tierra del Fuego (shared between present-day Argentina and Chile)
Ecuador (specifically Galápagos Islands)
French Polynesia (Tahiti)
New Zealand
Australia
Cocos (Keeling) Islands (territory of Australia)
South Africa
Ascension Island (British Overseas Territory)
Saint Helena (British Overseas Territory)

Portugal (specifically Azores, an archipelago in the mid-Atlantic)

Now, of course, much of that time would have been spent in transit, nonetheless much of that time would have been spent *observing* and having *continual exposure* to all the specimens and samples he was documenting in each of the regions he travelled to.

Furthermore, Darwin spent a period of *five weeks* alone at the Galapagos Islands studying the finches from which many of his key insights were obtained. Darwin is known to have initially noticed the finches during his exploration of the Galápagos Islands by foot. It is worth noting that Darwin was not specifically seeking out finches; rather, they caught his attention due to their diversity and the variations he observed among them. Despite the proximity of the islands to each other, Darwin was baffled by the variances he was seeing with the finches' physical characteristics from colourage, markings, feather patterns and then, the big one (as we shall see) – the beaks. First studying each of the finches and then specifically each of the beak shapes and sizes, Darwin *attempted* to create an explanation for why there was so much variance between them. One explanation Darwin considered was the idea of separate creations, which proposed that each species was independently created in its current form and placed in its specific environment by a divine creator.

Another idea that Darwin entertained was the possibility that the variations in beak shapes were the result of environmental influences acting directly on the organisms during their lifetime. This idea, known as Lamarckism[35] or the inheritance of acquired characteristics, suggested that organisms could pass on traits acquired during their lifetimes to their offspring.

However these theories failed to explain what he was observing and let us examine the reasons why. The first theory proposed that there was a divine creator: this explanation was dispelled, through Darwin's observations of the geographical distribution of species. He would often observe major *differences* between species found in nearby regions but *similarities* between species in faraway regions from one another.

For example, he observed greater similarities between mockingbirds on the Galápagos Islands and mainland South America at times, then of mockingbirds on different islands of the Galápagos archipelago (the Galapagos Islands and mainland South America are almost 1000 kilometres apart from each other). This appeared inconsistent to Darwin and it seemed quite disorderly for a creator to be placing species with similarities so far away from each other and dissimilar species so close. Though the idea of divine creation was definitely plausible for Darwin at the time, the differences he was observing *did seem* to have some kind of

[35] A canonical example used to explain Lamarckism is the giraffe's neck. Lamarckism suggests that an earlier version of a giraffe reached for a leaf at some point and as a result stretched or extended its neck ever so slightly in its lifetime. This extended neck would be passed to its offspring which also perhaps extended the neck reaching for some higher leaf and so on, until we have the giraffe of the present day with incredibly long necks.

pattern or regularity to them that did not appear to be attributable *necessarily* to a creator.

Secondly there was the argument that changes in species characteristics occurred *during* the species lifetime and were then inherited by their offspring immediately. This hypothesis, known as Lamarckism[36], was also rejected by Darwin. One of the main reasons for his rejection was that Lamarckism could not explain how complex and intricate structures such as an eye or skin develop. How could such complex external features develop when a species presumably has no control over such visual features (especially their outer physical features such as wing colour, skin colour, fur density). If a bird cannot change the colour of its wings or the design of its eye, how then could a bird species with newly coloured wings or eyes develop by Lamarckism?

So from the various explanations, if we look closely, we can observe that these explanations function as the 'attempts' Darwin was making at creating the theory of natural selection and the explanations we have covered are only two potential explanations (divine creation and Lamarckism) he is known to have considered. Who knows how many attempts of different explanations Darwin

---

[36] Lamarckism is the theory that traits an organism acquires during its life could be passed on to its offspring.

For example:

A giraffe stretches its neck to reach high leaves.

Over generations, this stretching leads to longer necks in descendants.

The core of this theory is that traits that are used more become stronger; traits that are not used weaken and may disappear.

actually came up with before arriving at natural selection. That too Darwin arrived at the Galapagos Islands almost four years into his journey *after* making various observations of flora, fauna and geological specimens before even arriving at the islands. This again reiterates just the length of continuous exposure that Darwin had to the different environments he was observing during his voyage. This in our terms, means that Darwin had *high familiarity* with the natural environment of a number of different geographical regions, which was definitely *sufficient* enough to begin making *attempts* at explaining natural selection.

So we are at the stage now where Darwin has sufficient familiarity and has begun creating potential explanations for his observations which included the Divine Creator hypothesis and Lamarckism. He now needs the *progressive link* to create the final explanation for natural selection which happens when Darwin draws a key insight upon a certain kind of bird and a very important physical characteristic - the finches and their beaks.

Darwin is known to have collected a variety of finches during his five day exploration of the San Cristobal island which he examined once he had returned to England. This island was a huge component in Darwin's eventual explanation for natural selection since it eliminated the parameter of geography. In all his other observations, Darwin was attempting to draw connections between species on different islands and environments. Now Darwin was confined to one geographical location; and despite this isolation,

Darwin was still observing significant differences between the finches on the island. It was at this point, through his observations, Darwin noticed the key connection that catalysed his formulation of natural selection - *the beak shapes of the finches were correlated with the different food sources they consumed on San Cristobal island.* Putting this connection in the scope of our discussions on numerical intuition, what happened here was that Darwin saw the *progressive link* between the physical features he was familiar with (namely the finches beaks) and the food sources they were consuming.

Additionally from Darwin's perspective, it wasn't just that the beaks were *correlated* with the different food sources, it just so happened that the finches beaks were *perfectly adapted* to consuming their respective food sources, having the relevant shape and size for optimal consumption. Darwin now saw that the *environment* actually played a role in the characteristics a species exhibited and this is the most central idea of the final theory of natural selection. In a similar fashion, Darwin then began to see the links between more complex components of a species such as behavioural adaptations, reproductive variances and feather colourings and the central idea of the *environment* playing a factor into these adaptations.

It must be said however even after all of these additional connections Darwin was finding, the observation of the food sources and their correlation with the beak shapes and sizes, was pivotal since it drew the *first* deep connection with the environment and as Darwin put it 'the Origin of Species' (the unique characteristics of each species).

So what we have is that Darwin progressed through numerous stages of *familiarisation* including the finches, the beaks, the environment etc and through his repeated attempts at explanation, he was able to find the connection between each of the components. Through familiarity Darwin was then able to form a comprehensive explanation to the origin of species – without understanding any of the physical mechanisms that drive evolution at the genetic level.

Sufficient Familiarity: Dmitri Mendeleev [16]

Along with Charles Darwin, another one of the most famous examples throughout history of progression through familiarity is that of the Russian chemist Dmitri Mendeleev (1834 – 1907), who invented the periodic table (not in the form that we know of it today but the most rudimentary form before its extension by later chemists).

Dmitri Mendeleev (1834 – 1907): Russian chemist who created the periodic table of elements, organizing them by atomic weight and properties, which predicted the existence of undiscovered elements.

Mendeleev's life represents a particularly interesting case for familiarity because scientists at that time knew practically nothing about the most fundamental constituent of the substances they were examining - the atom. The understanding of the atom and chemical properties of various elements during Mendeleev's lifetime pales in comparison to what we know today. Admittedly, this can be said of virtually any field but it is particularly pertinent within the field of chemistry. As a field, chemistry is all about classifying substances based on their physical properties and in Mendeleev's lifetime, even knowing what these substances were made of was a subject of debate and was limited to the various theorems of that time period[37].

Additionally, many of the underlying patterns, structures and theoretic formulations underpinning the periodic table were developed much after Mendeleev's life – yet he was still able to develop the periodic table through sufficient familiarity.

[37] The main one at the time being English physicist John Dalton (1766 - 1844). Dalton proposed that atoms are indivisible, solid spheres and each element is composed of identical atoms. While Dalton was correct that each element is composed of individual atoms, we know now that atoms are not solid spheres but mostly empty space.

Short Biographical Account of Dmitri Mendeleev

Dmitri Ivanovich Mendeleev, one of the most influential figures in the history of chemistry, was born on 8 February 1834 (27 January O.S.) in the Siberian town of Tobolsk, Russia. He was the youngest of a large family, often said to be the fourteenth child, though records are uncertain. His father, a teacher, lost his eyesight and died when Dmitri was still young, leaving the family in financial hardship. His mother managed a glass factory to support her children and placed great emphasis on education, encouraging Dmitri's talents.

In 1850, Mendeleev entered the Main Pedagogical Institute in St. Petersburg, a school for training teachers, where he specialized in natural sciences. Despite suffering from ill health and poverty, he excelled in his studies and graduated in 1855. His early research earned him recognition, and in 1856 he completed a master's degree in chemistry at St. Petersburg University with a thesis on isomorphism and the relationship between chemical composition and physical properties.

After teaching briefly in St. Petersburg and Simferopol, Mendeleev traveled abroad for further study. From 1859 to 1861, he lived in Heidelberg, Germany, where he worked independently on the physical properties of liquids and their relation to molecular forces. While in Heidelberg he attended the Karlsruhe Congress of Chemists (1860), a landmark meeting where Stanislao

Cannizzaro's clear presentation of atomic weights deeply impressed him. This encounter would later prove vital in shaping his ideas on atomic theory and chemical periodicity.

Returning to Russia in 1861, Mendeleev began teaching and quickly built a reputation as a gifted lecturer and researcher. He defended his doctoral dissertation in 1865, titled On the Combinations of Water with Alcohol, which examined the unusual properties of solutions. The following years saw the publication of his textbook, Principles of Chemistry (1868–1870), which systematically presented chemical knowledge and emphasized relationships among the elements.

It was while preparing this textbook that Mendeleev achieved his greatest breakthrough. In 1869, he formulated the Periodic Law, arranging the known chemical elements in order of increasing atomic weight while recognizing repeating patterns in their properties. His Periodic Table not only organized existing knowledge but also boldly predicted the existence and properties of elements yet to be discovered, such as gallium, scandium, and germanium. When these elements were found and matched his predictions closely, Mendeleev's system gained worldwide acceptance and became the foundation of modern chemistry.

Beyond the periodic table, Mendeleev contributed to many areas of science and technology. He studied the behavior of gases and liquids, explored the structure of petroleum, and promoted the development of Russia's oil industry. He also worked on agricultural chemistry, the design of smokeless gunpowder, and standardization of industrial processes. In 1893, he was appointed Director of the Bureau of Weights and Measures, where he championed the adoption of the metric system in Russia.

Mendeleev's personality was complex: he was imaginative, independent, sometimes combative, but always deeply committed to science and education. He resigned from his university post in 1890 after disputes with authorities, but his scientific influence remained strong.

Dmitri Mendeleev died on 2 February 1907 (20 January O.S.) in St. Petersburg at the age of 72. Remembered above all for the Periodic Table, his legacy extends far beyond a single discovery, embodying the systematic spirit of modern science.

Mendeleev and Sufficient Familiarity

One has to wonder how a chemist with such little information on atomic theory could have developed such a remarkably accurate invention. Let us break down some of the key components of Mendeleev's development of the periodic table to see how familiarity played a central role. The first point to consider is that of Mendeleev's day to day profession - Mendeleev worked as a lecturer and professor at the University of St Petersburg where in addition to his professorship based duties, he performed a variety of experiments where he was able to determine the various properties of the elements he knew of at the time (atomic mass, valency, volume and chemical reactivity). This amounted to a large body of data that Mendeleev actively documented and revised upon the completion of each successive experiment. What we can observe here is that Mendeleev went through a prolonged period of continual exposure to the elements he was studying through all of his experiments. Furthermore, Mendeleev was known to have performed experiments almost daily, analysing properties of not just the chemicals on their own (which is actually very hard for reactive elements such as Sodium), but their compounds too (sodium hydroxide etc).

Mendeleev then began trying to arrange the elements in various ways to see if there was any pattern amongst them and their properties. Mendeleev first began arranging by atomic volume[38].

[38] Atomic volume is a measure of how much space an element occupies. Mendeleev would measure this by directly calculation (for example if the element had a cylindrical or regular shape, common formulas

He observed that certain groups of elements exhibited similar chemical properties when arranged according to their atomic volumes. However, this approach did not yield a satisfactory periodic pattern since atomic volume does not exhibit a clear, uniform trend across the elements.

Mendeleev then experimented with grouping elements into triads and octaves based on similarities in their properties. He observed that certain sets of three elements or groups of eight elements exhibited recurring patterns. While these groupings hinted at the periodic nature of the elements, they were not as systematic or comprehensive as the whole of the periodic table.

Mendeleev also considered organizing the elements according to their valence, or the number of bonds an element can form with other atoms. He observed that elements with similar valences often exhibited similar chemical behaviors. While valence played a significant role in the final arrangement of the periodic table, it was not the sole criterion used by Mendeleev.

Ultimately, Mendeleev settled on arranging the elements in order of increasing atomic weight as the basis for his periodic table. This approach allowed him to identify the periodic patterns in the chemical properties of the elements, leading to the development of the periodic law and the modern periodic table.

---

could be applied to find volume) or the water displacement method (Archimedes principle) where he would submerge the solid element in water and measure the displayed volume.

What we can then observe is that these attempts at arrangement of the elements constitute Mendeleev’s stage of performing repeated computations/attempts in finding the proper arrangement of the elements. All the while Mendeleev was developing and progressing his understanding with regards to any possible relations between the elements, until eventually arriving at the arrangement that was best. Mendeleev's exploration of different arrangements was an iterative process, with each approach informing the next. For example, his observations of recurring patterns in triads and octaves influenced his decision to organize the elements by atomic weight, as he recognized the periodic nature of these patterns. In this way by developing familiarity with one arrangement Mendeleev then drew a connection to the next possible arrangement. Each one of the potential arrangements encoded a piece of information Mendeleev could then use to build a more advanced arrangement and progress his understanding of the arrangement of the elements until he finally had one that worked.

Predictive Capabilities Arising from Sufficient Familiarity

Mendeleev's case is interesting because it not only shows progression through sufficient familiarity but also how familiarity alone can give some predictive capacity within a field. Within his initial arrangement of the periodic table Mendeleev correctly predicted the existence of various elements and their different properties including their atomic weights, chemical and physical properties (and even names for the new elements).

Some of Mendeleev's most famous predictions included the elements gallium, scandium, and germanium, all of which were discovered after Mendeleev's publication of the periodic table and the properties of which were found to closely match his predictions. These successful predictions provided strong evidence for the validity of Mendeleev's periodic table and highlighted its predictive power in guiding future discoveries in chemistry.

Mendeleev's case demonstrates how a practitioner, despite not having understanding of the underlying principles at play, can still develop a feel for how things should work within a field. What's remarkable also is that a proper model of the atom was not even constructed at the time of Mendeleev's publications. Better yet the chemical properties of an element, that Mendeleev was predicting is now known to be highly linked to the number of valence electrons it has and electrons were not even discovered until 1897 (almost 30 years after his first publication on the periodic table)!

To emphasise how little Mendeleev actually knew about the atom let us consider for a moment what *was* actually known about the atom at the time.

Indivisibility: Mendeleev was aware of the prevailing belief in the indivisibility of the atom, which was proposed by early Greek philosophers such as Democritus. According to this idea, atoms were considered to be the smallest, indivisible units of matter.

Law of Multiple Proportions: this law which states that when elements combine to form compounds, the ratios of the masses combine in a fixed manner and can be expressed in small whole numbers. This empirical law provided evidence for the existence of atoms as discrete, indivisible entities.

Mendeleev recognized the concept of atomic weight as a fundamental property of the atom. He understood that the atomic weight of an element represented the mass of its atoms relative to the mass of a standard reference atom (like hydrogen).

Mendeleev understood that the chemical properties of elements were determined by the structure of their atoms and the interactions between atoms during chemical reactions. His periodic table was based on the observation that elements with similar chemical properties tended to occur at regular intervals when arranged by atomic weight.

While Mendeleev did not have detailed knowledge of isotopes, he was aware that some elements existed in multiple forms with different atomic weights. This understanding influenced his decision to leave gaps in his periodic table for undiscovered elements and to predict their properties based on their positions relative to neighbouring elements.

Yet the patterns, the computational methods and measurable properties Mendeleev was aware of served as a basis for Mendeleev to still identify what the missing elements' properties should be. Mendeleev's case actually draws many parallels within mathematics. His familiarity with elements is very similar to the familiarity a practitioner develops working with certain rules for integration and derivatives etc. Much like Mendeleev a practitioner can have absolutely no idea of the underlying axioms and proofs that govern why they can even perform certain mathematical operations, nonetheless an intuition is still developed for the performance of the operations much like how Mendeleev had an intuition for the position of yet-to-be discovered elements.

Sufficient Familiarity Case: Isaac Newton [17]

Now I know what you're thinking - surely not Isaac Newton? The father of modern gravity (the grandfather perhaps if you include Einstein) surely had a full understanding of the theorems he was proposing at the time of publication right?

This is far from the truth, and the early 1600 – 1700s mathematicians themselves are very interesting to study when it comes to the development of their theorems. What is usually forgotten about this period of time is that many of our currently established scientific practices had not yet been proposed. It's one thing to talk about a theorem once it has been published: to critique it, to test it or to extend upon an already existing understanding is very different from actually developing the understanding in the first place. To see this, I encourage the reader to try developing their own theory of planetary motion, that's accurate and allows them to predict where objects will land if thrown, or where Neptune will be positioned in September 2055. After five minutes, a practitioner will begin to see that it is very unclear on where to even begin, so one must take this perspective on novelty forward when analysing the 1600 - 1700s mathematicians and their contributions at a time where there was no prior developed framework. Furthermore, many of our current methods for performing empirical observations did not exist in this time period, so even if a practitioner did develop some groundbreaking theorem it could be difficult to test and verify[39]. Keep in mind it was only around Newton's time that telescopes and pendulum clocks for time based measurements were even invented.

Isaac Newton (1642 - 1727): English mathematician, physicist, and astronomer who made foundational contributions to science through his laws of motion, and contributions to the theory of gravitation.

[39] Case in point are Kepler's Laws of Planetary Motion the first two of which were published in 1609. Kepler's laws were experimentally verified to be true mostly through the use of the telescope and the first known telescopes were built between 1608-1609 (by Galileo and others) in the very same year of Kepler's publications. To verify Kepler's laws astronomers essentially had to wait until they had enough data from the telescopes to back up his laws proposals[18].

Without being able to directly observe the very objects one is making predictions for, it can be extremely difficult to craft a theorem around their motion as one would suspect. Thus what we see with Newton as with many other philosophers of his time, is that their theorems *begin* more as a philosophical framework that is designed to correspond to real life observations. Further experimental evidence then substantiates their philosophical claim, which then morphs the philosophy into a kind of real world understanding. Nonetheless the ideas all begin as a philosophical postulation. To emphasise this point - did you know that Leibniz also had his own set of theorems for the motion of bodies?

One of Leibniz's key contributions to mechanics was the concept of "vis viva," or "living force", which is the precursor to what we now refer to as kinetic energy. Leibniz proposed that the motion of an object is determined not only by its velocity (v) but also by its mass (m). He argued that the quantity of motion (vis viva) possessed by an object is proportional to its mass and the square of its velocity, mv^2 (this proposal is very close to the kinetic energy equation we have today).

Now no one in the current field of science talks about 'vis viva' (living force) when referring to kinetic energy and it sounds, well, quite outlandish to be frank. However in Leibniz's day one must realise that there was *no* theorem behind the motion of bodies and to call something 'living force' or 'live movement' or 'living potential', because of a fundamental *philosophical* claim[40] that objects possess such inherent characteristics, is maybe not as eccentric as it might seem. Had Newton called force, 'vis viva' maybe everyone would also be giving it the same name but since Leibniz's framework did not popularise, 'force', as Newton described it in his publications became the prevailing term.

[40] For example, Leibniz believed that the fundamental units of reality are *monads*, which are simple, immaterial, indivisible substances. Monads are the building blocks of all things, but they do not interact directly with each other. When you already believe objects to be composed of an immaterial, indivisible substance if you also claim that they have an inherent potential for motion (or something along these lines) it's easy to see how you can go from monads to something like 'vis viva' (living force). I am not claiming this is how Leibniz proposed 'vis viva', but within such philosophical domains it's understandable how such terms for physical elements of reality can arise.

Short Biographical Account of Isaac Newton

Isaac Newton, one of the most influential scientists in history, was born on December 25, 1642, in Woolsthorpe, Lincolnshire, England. His father, also named Isaac Newton, had passed away three months prior to his birth. Newton's early years were marked by hardship and uncertainty, but he displayed exceptional intellectual abilities from a young age.
Newton attended the King's School in Grantham and later enrolled at Trinity College, Cambridge, in 1661. At Cambridge, he studied mathematics, natural philosophy, and optics, laying the groundwork for his future groundbreaking discoveries. Despite facing financial struggles, Newton immersed himself in his studies and excelled academically.

During his time at Cambridge, Newton began formulating his revolutionary theories on calculus, optics, and the laws of motion. He made significant advancements in mathematics, developing calculus independently of Gottfried Wilhelm Leibniz, and laying the foundations of what would later become known as classical mechanics.

In 1665, the outbreak of the bubonic plague forced Cambridge University to close its doors, prompting Newton to return to his family estate in Woolsthorpe. It was during this period of isolation, known as his "annus mirabilis" or "miraculous year," that Newton made several groundbreaking discoveries. He formulated his laws of motion, developed the universal law of gravitation, and made significant progress in optics, including his experiments with prisms and the nature of light.

In 1667, Newton returned to Cambridge and was elected a Fellow of Trinity College. He began lecturing on mathematics and optics, attracting widespread recognition for his intellect and innovative ideas. Newton's reputation continued to grow, and in 1669, he was appointed Lucasian Professor of Mathematics at Cambridge, a prestigious position he held for over three decades.

In 1687, Newton published his seminal work, "Philosophiæ Naturalis Principia Mathematica" (Mathematical Principles of Natural Philosophy), commonly known as the Principia. In this monumental work, Newton presented his laws of motion and the law of universal gravitation, providing a comprehensive mathematical framework for understanding the motion of objects and the forces acting upon them. The Principia revolutionized physics and laid the foundation for classical mechanics.

Throughout his life, Newton made significant contributions to various fields of science and mathematics. He conducted experiments in optics, developed theories on the nature of light and color, and made pioneering contributions to the study of calculus and mathematics. In addition to his scientific work, Newton was also involved in alchemy, theology, and biblical studies, reflecting the breadth of his intellectual interests.

Isaac Newton's legacy extends far beyond his scientific achievements. He was knighted by Queen Anne in 1705 and served as President of the Royal Society from 1703 until his death in 1727. Newton's work laid the groundwork for modern physics and mathematics, and his principles continue to shape our understanding of the universe. He remains a towering figure in the history of science, revered for his intellect, ingenuity, and profound contributions to human knowledge.

Newton's Law of Gravitation

Though the cases of Darwin and Mendeleev are interesting in their own right, the case for Newton is distinct from the previous two. This is because Isaac Newton firstly acted as more of a philosopher than what Darwin and Mendeleev did. Newton formulated a framework to explain the motion of bodies using only the information he could attain from empirical observations and the mathematical tools available to him at the time. Beginning with the laws of motion, Newton applied these laws to explain the concept of gravity which he believed is a fundamental force of nature that acts between all objects with mass. This concept of gravity was then mathematically formalised (made into an equation or measurable concept) in the universal law of gravitation which is written as below:

$$F = G\frac{m_1 m_2}{r^2}$$

The reasoning he applied within this derivation, came only from the laws of motion
which he had formulated. An outline of these laws is given below which were included within his magnum opus - *Principia*:

1) Lemma I: In this lemma, Newton establishes the concept of centripetal forces acting on bodies moving in curved paths. He shows that if a body moves uniformly in a circular path,

it is acted upon by a centripetal force directed toward the center of the circle (F).

2) Lemma II: Newton demonstrates that if a body moves uniformly in an orbit around a central body under the influence of an attractive force directed toward that body, then the force of attraction varies inversely with the square of the distance between the two bodies (r^2).
3) Lemma III: In this lemma, Newton proves that the force exerted by a uniform spherical shell on a point mass outside the shell is equivalent to the force of a point mass located at the center of the shell (this allows one to apply gravity to spherical objects, like planets, as opposed to point masses).
4) Lemma IV: Newton shows that the force exerted by a uniform spherical shell on a point mass inside the shell is zero ($F_{net} = 0$ on both points of mass).
5) Lemma V: This lemma deals with the gravitational forces exerted by two bodies on each other when they are not spherical. Newton establishes that if the distance between the centers of two such bodies is sufficiently large compared to their dimensions, then the force between them can be treated as if their mass were concentrated at their respective centers. (m_1 and m_2).

By leveraging these lemmas and combining them with his laws of motion, Newton was able to rigorously derive the law of universal gravitation. This was a monumental achievement in physics and laid the foundation for our understanding of gravity and he was able to complete this derivation, simply from familiarity with the lemmas above and his ability to compute (reason) with them.

It is also interesting to note that in his lifetime, Newton himself admitted that he did not fully understand the mechanism by which gravity operated - he famously wrote in a letter to Richard Bentley[41] in 1692: "That one body may act upon another at a distance through a vacuum without the mediation of anything else, by and through which their action and force may be conveyed from one to another, is to me so great an absurdity that, I believe, no man who has in philosophic matters a competent faculty of thinking could ever fall into it." This statement reflects Newton's struggle to reconcile his gravitation with his intuition about the nature of physical interactions - he was astonished that *his* laws of motion could lead to such counter-intuitive conclusions.

---

[41] Richard Bentley (1662–1742) was an English scholar, theologian, and classical critic, widely regarded as one of the most brilliant intellectuals of the early 18th century. He is best known for his work in classical literature, textual criticism, and theology. Yes, you read that write 'textual criticism' - that means to look at a piece of text and critique whether its contents are the same as the original document. Bentley applied this 'textual criticism' on a number of works in the Western canon including the *Iliad, Odyssey* and the *New Testament.*

Newton and Sufficient Familiarity

So what can we observe in the case of Newton and his relation to sufficient familiarity?

To me this is best explained through the following analogy which actually has nothing to do with physics and everything to do with cooking (just believe me this will explain how Newton is a prime case of sufficient familiarity):

Cooking Analogy: Whenever you start cooking you begin with a set number of ingredients. A good cook will have an idea of a number of different recipes that can be created from the ingredients. Now some of these recipes might be grand ones such as a special pasta dish or lesser known ones like a cheesecake but all are recipes nonetheless. The creation of any of these recipes is dependent, on a high part, by the familiarity the chef has with the ingredients and their ability to cook with them.

Now I am going to replace the following words in the *cooking analogy* with the words below, to create the *Reasoning Analogy* and the reader will see exactly how the analogy relates to Newton:

cook (verb) = compute
ingredients = objects
cook (noun) = philosopher
recipe = theorem

special pasta dish = Law of Gravitation

cheesecake = Newton's method

Reasoning Analogy: Whenever you start *computing* you begin with a set number of *objects.* A good *philosopher* will have an idea of a number of different *theorems* that can be created from the *objects.* Now some of these *theorems* might be grand ones such as the *Law of Gravitation* or lesser known ones like *Newton's method (for approximation)* but all are *theorems* nonetheless. The creation of any of these *theorems* is dependent, on a high part, by the familiarity the *philosopher* has with the *objects* and their ability to *compute (reason)* with them.

Hopefully this analogy helps to clarify some of the works of Newton and his relation to familiarity before we begin with the mathematical component as below.

The basis of a majority of Newton's derivations arise from his laws of motion, which through a combination of empirical observations, experiments and the formalisation of previously introduced concepts by Galileo, he was able to derive. Now having derived these laws from deep contemplation on the potential principles governing the motion of bodies, Newton had a strong degree of familiarity with the laws and the areas where these laws could be applied to to understand the motion of bodies.

This was not the only component though, another huge component of Newton's work is his application of calculus to the derivations to really *formalise* his work in motion. Prior to the application of calculus, his laws really just exist as empty words and to any practitioner this is exactly how they would first appear (certainly did for me). Now not only did Newton apply calculus in the formulation of his laws, he *invented* calculus[42]. In fact he even outlines the ideas behind calculus in his text, published after his death, 'The Method of Fluxions and Infinite Series' which is over 300 pages long.

Now if a practitioner has written a book about a subject we can conclude that they will (most likely) have high familiarity with any of its key concepts no matter how looney they may seem. We really have to take a moment to note just how outlandish some of his laws *initially* sound without having some rigorous mathematical formalism to justify them. Here is an example of his law of inertia which we define as the tendency of an object to resist a change in its motion. Now what is written below is actually from a translated version of his Principia:

Vis Insita (Inherent Force): The vis insita, or innate force of matter, is a power of resisting by which every body, as much as in it lies, endeavors to preserve its present state, whether it be of rest or of moving uniformly forward in a straight line.

[42] Yes I understand, Leibniz also invented it concurrently and there is a minor debate to this day about who invented it first. An interesting note is that this debate was happening even during their lifetimes - Leibniz even had his own personal qualms with the Royal Society in London.

You may have noticed something with the definition above - the word 'inertia' isn't in it.

You can see instead the latin Vis Inhita which is described by Newton as the 'innate force of matter' and the fact that a body 'endeavours to preserve its present state' (as though an inanimate object has the ability to 'endeavour'). It can be humorous at times to see the number of physics students, or university professors that use the term 'inertia' like it's an obvious part of nature, and have never understood how Newton actually referred to it.

But what happens when you throw a little bit of calculus on the same definition?

Well from the definition we need a force to overcome this Vis Inhata, let's call it F and we can say that mass, m, is somehow related to inertia as more massive bodies have a greater tendency to resist changes in motion (it's harder to move a your car than a feather). So we know that F and m are connected in some way. We also then need a way to get the body to move, so a force needs to act on the body causing some change in its velocity for it to actually move. Let's call this change in velocity, acceleration and bingo you get Newton's Second Law of Motion:

Newton's Second Law of Motion: The force acting on an object is equal to the mass of that object multiplied by its acceleration.

$$F = ma$$

Now that was a huge simplification of the derivation, especially that last step we just took, but now consider how we were able to perform this derivation.

We were familiar with the law of Vis Inhita, which we ourselves came up with, we understood that mass was connected with the concept of Vis Inhita. We then applied calculus, which we also had a high familiarity with from presumably having invented it and in the end we arrived at a simple equation (F = ma). Now there are many fine steps of reasoning required to arrive at the formula, but the basis of it, as we can see through this exercise, is the familiarity we had with each key component of the principles applied, whether it was mass, calculus, Vis Inhata or force. Through the confluence of these concepts (in a way performing computations with them), we then are able to reach the final result.

Now the reader may be thinking that perhaps it was just *this* equation shown in this text, $F = ma$ or the Law of Gravitation that Newton was applying the principles of familiarity to derive. Well, let me remove your confusion. These base principles were applied in *all* of Newton's derivations.

Part 4: Properties

Rene Descartes (1596 - 1650): "Divide each difficulty into as many parts as is feasible and necessary to resolve it."

Introduction to Properties

Almost every field has some concept which is perceived as 'hard' or 'difficult'. Whether it's linear algebra in mathematics, heart surgery in medicine or tax law within legal studies, every field will have some area that practitioners tend to avoid that almost has a mythical lustre surrounding it. Anyone brave enough to venture into such fields will be perceived as, at times, crazy or insane or just plain courageous. But what makes these areas difficult for us to understand and work with?

For example, we can certainly say that mathematics and law are different fields from the outset - one involves communicating with clients, the resolution of ethical matters and metaphysical concepts of morality and justice where the other pertains mostly to computation and logical reasoning. These are completely different fields in content yet both are perceived to be difficult. So we can then ask, is there a deeper reason that entails the difficulty of these fields rather than their subject matters?

Within mathematics there are distinct *properties* of the field that intrinsically differentiates it from all others, and we will analyse some of these properties in detail to see the importance of numerical intuition and sufficient familiarity. The reader thus far may be led to believe that numerical intuition and sufficient familiarity are just creations of the author and just happen to be applicable to mathematics. However this is not the case - the concepts of numerical intuition and sufficient familiarity aren't arbitrary concepts that *happen* to be applicable to mathematical learning - there exist deeper reasons why we can apply them to mathematics.

To give a brief example to the reader, consider for a moment the following question:

Is it necessary to learn one song on piano before learning another?

Most people, I would imagine, would answer with 'no', and we will come back to this example in later sections. But if we were to apply the same kind of question within mathematics:

Is it necessary to learn one area of mathematics before learning another?

Most people, I would imagine would say 'Yes', a quick example would be addition before multiplication or to teach shapes and lines before teaching what angles are to a practitioner.

From a simple question we can already see a stark difference between mathematics and music as fields and a natural question would then be the following:

What properties of mathematics make it a distinctive field of study?

Within the below excerpt we will analyse three key properties, that make mathematics distinctive which have a strong connection to sufficient familiarity and numerical intuition:

1) *Precedence: the property of a field where one requires familiarity with a concept to work with (perform computations) other concepts.*
2) *Intrinsic Difficulty: the property of a field where problems generally require a high number of elementary steps to solve.*
3) *Interdependence: the property of a field where most concepts are related to one another.*

Property 1: Precedence

To begin this section let's first look at the definition of precedence again which was introduced above:

Precedence: the property of a field where one requires familiarity with a concept, to work with (perform computations) other concepts.

So what exactly do we mean here?
In very simple terms it's something along the lines of: 'You have to know how to walk before you know how to run'- or rather that one requires familiarity with walking (one concept) to be able to run (other concept).

It would appear obvious that this property should exist in every field of learning. Surely in every field one must be familiar with concepts that precede others as this could only be the logical direction for progression right?

To examine why this is property of mathematics and *not of other fields* let us actually consider how one *learns* in a number of other fields. One classic example we can draw from is the field of music, specifically playing an instrument. For the purposes of this analysis let us consider the example of playing the piano. In learning the piano, a practitioner begins learning simple pieces of music with basic chords, to improve hand dexterity and coordination. This is all done with the aim that the more advanced pieces would become easier to play as they *require* greater dexterity and hand coordination to play than the simpler pieces.

However there is a key distinction between a piece of music becoming *easier* to play and a piece of music being *possible* to play. Even without learning simpler pieces of music it is hypothetically possible that one can 'skip' the easier pieces, and *progress* to an advanced piece just by learning the necessary keys of the advanced piece. Now if a practitioner dedicated all of their time to learning a single advanced piece of music then their skills would be limited to *that one* advanced piece of music. It would also take the practitioner much longer to properly play the advanced piece without having first learnt any simpler pieces of music – nonetheless a practitioner *can theoretically* still play the advanced piece if given enough time.

To give another basic example of a field that demonstrates precedence, we can also consider the field of reading. When a practitioner learns how to read they first learn the letters of their languages alphabet. Then the practitioner will start to read words and eventually, full sentences. In an astonishingly short amount of time a practitioner can then begin to read books that use simple language and eventually books using more complex language.

From the moment one learns to read, a practitioner *can* hypothetically read a complex text such as '*To kill a Mockingbird*' by Harper Lee. Sure, the practitioner may not understand or even know how to pronounce each word, but they *can* still *read* each word - and by this I mean that they can still recognise each letter in a word and know that it forms a single unit of meaning. Knowing how to pronounce this unit or what this unit means is a completely separate area which we will discuss in *Numerical Intuition: Understanding*. Furthermore one does not need familiarity with the entire English language to comprehend the book - only those that actually appear in the book suffice. Thus if it's true that a complex text is readable from the moment one learns to read (though they may not understand the text) we can observe that reading simple texts is not actually required to be able to read the advanced ones by extension.

Now contrast playing the piano and reading with the field of mathematics where, to even learn any new concept, familiarity is required with key concepts which *precede* the introduction of the new concept. For example in mathematics we usually begin with learning numbers, counting and basic arithmetic with numbers before more difficult operations such as multiplication and division and then fractions and so on.

When you get to something like multiplications it is easy to see the basic fact that we cannot multiply numbers if we do not know how to first add them. We can observe that a practitioner must be familiar with a prior concept (addition) before being able to perform *computations* in the more advanced concepts (multiplication).

Now that we have established what precedence is, through a simple example, we can then begin to see that mathematics as a field *follows precedence.* Mathematics is not like other fields such as computer science where one can specialise in a particular algorithm or field of research with absolutely no knowledge of any preceding concept. Whether its topology, vector calculus, matrices – basic operations are always performed such as addition, integrals, subtraction etc on any of the objects within these fields of study. Lacking knowledge of the base operations within a field – *computations* – necessarily means there can be no development within numerical intuition in a related field which relies on these base operations, as one cannot perform the very computations required to develop numerical intuition!

Let us now look at an actual example within mathematics of the precedence property, to give greater context to the ideas we are discussing.

Precedence Case: Matrix Multiplication Algorithms

Basics of Vectors

In this section we will focus on vectors and scalars (concepts) and how they *precede* matrices (related concept). Let us begin this section by first explaining what a scalar is. In mathematics a scalar is a numerical value that represents a magnitude. For example:

5 is a scalar
87 is a scalar
23 is a scalar

Now vectors (in 2 dimensions) are basically a pair of scalars, where the first number represents how horizontal the vector is and the second number refers to how vertical a number is. To give an example:

(1,2) is a vector
(18,9) is vector
(23,0) is a vector

We can give a geometric representation to a vector by actually drawing out the vector:

(1,2) : 1 horizontal and 2 vertical -

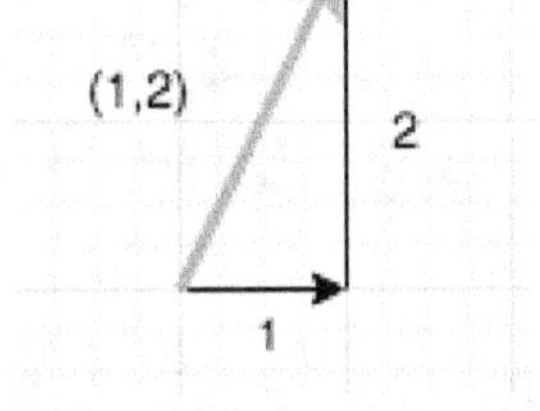

That green arrow shown in the image above is the vector (1,2) which has a value of one for its horizontal component and two for its vertical component. If we wanted to scale this vector by a factor of 3, we multiply each of its components by 3, which would give the vector (3,6) (3,6) compared with (1,2).

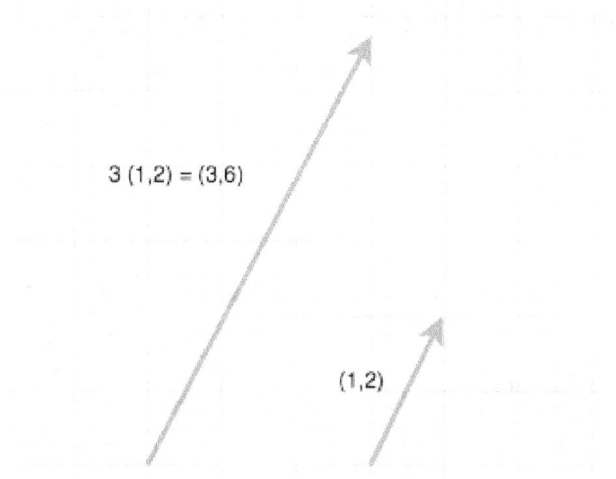

For the purposes of what we will discuss in this excerpt just the above background on vectors and scalars is *sufficient*. One of the first operations anyone learns with vectors, is called the dot product which is a way of turning two vectors into a scalar. It is defined like this:

(a,b) . (c,d) = ac + bd

So to give an example the dot product of the vectors (1,2) and (3,4) is:

$$(1,2) \cdot (3,4) = 1 \times 3 + 2 \times 4 = 11$$

We defined the dot product above for vectors of two dimensions but we can do the same operations in three, four, five or any number of dimensions. If we were to take the vectors (1,2,3) and (4,5,6) which are three dimensional, the operation works the same:

$$(1,2,3) \cdot (4,5,6) = 1 \times 4 + 2 \times 5 + 3 \times 6 = 4 + 10 + 18 = 32$$

This covers the basics of vectors and scalars. Let us now introduce what matrices are. In mathematical terms a matrix is simply an array of numbers which can have any number of rows and columns. For example:

$\begin{bmatrix} 1 & 2 & 3 \\ 10 & 4 & 5 \\ 12 & 2 & 1 \end{bmatrix}$ is a matrix

$[24]$ is a matrix

$\begin{bmatrix} 1 & 8 & 2 \end{bmatrix}$ is a matrix

Now in terms of vectors what we generally do to write them in matrix format is we have a row (or column) for each number in the pair. So for the vector (1,2) it can be written as either:

$$\begin{bmatrix} 1 \\ 2 \end{bmatrix} \text{ OR } \begin{bmatrix} 1 & 2 \end{bmatrix}$$

Either of these notations is fine for the vector (1,2) depending on the required task.

Matrix Multiplication

The reason we introduced vectors and dot products above is that matrix multiplication is defined by dot products on the rows and columns of the matrices. The matrices have to have specific dimensions to be multiplied, specifically if we have two matrices A and B, they can only be multiplied if the number of columns in matrix A is the same as the number of rows in matrix B. The resulting matrix will have the same number of rows as matrix A and the same number of columns as matrix B. So for example:

$$\begin{bmatrix} 1 & 2 \\ 3 & 4 \end{bmatrix} \cdot \begin{bmatrix} 5 \\ 6 \end{bmatrix} \text{ can be multiplied}$$

$$\begin{bmatrix} 1 & 2 & 3 \\ 4 & 5 & 6 \\ 7 & 8 & 9 \end{bmatrix} \cdot \begin{bmatrix} 3 & 4 & 3 \\ 2 & 2 & 1 \\ 5 & 4 & 3 \end{bmatrix} \text{ can be multiplied}$$

$$\begin{bmatrix} 1 & 2 \\ 2 & 1 \end{bmatrix} \cdot \begin{bmatrix} 7 & 8 \end{bmatrix} \text{ cannot be multiplied}$$

To quickly verify if two matrices can be multiplied we can apply a trick, I was taught by one of my professors. If we want to see for example if:

$$\begin{bmatrix} 1 & 2 \\ 3 & 4 \end{bmatrix} \cdot \begin{bmatrix} 8 \\ 7 \end{bmatrix} \quad \text{can be multiplied}$$

We draw the matrices as shown below and we have that the 'lengths' of the column number in Matrix A and the 'length' of the rows in B are the same so they can be multiplied:

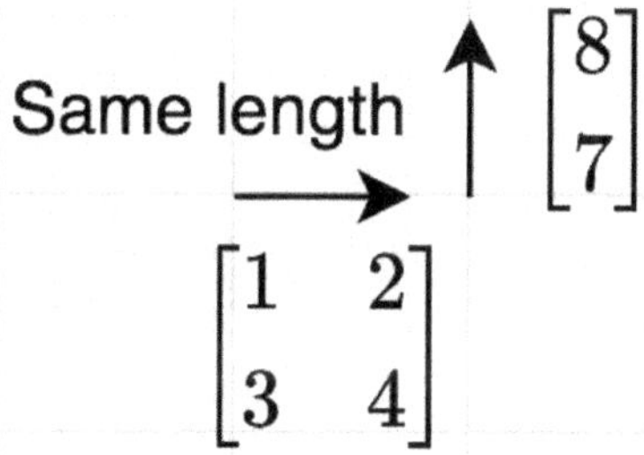

And the result will be the dot product of each row with each column:

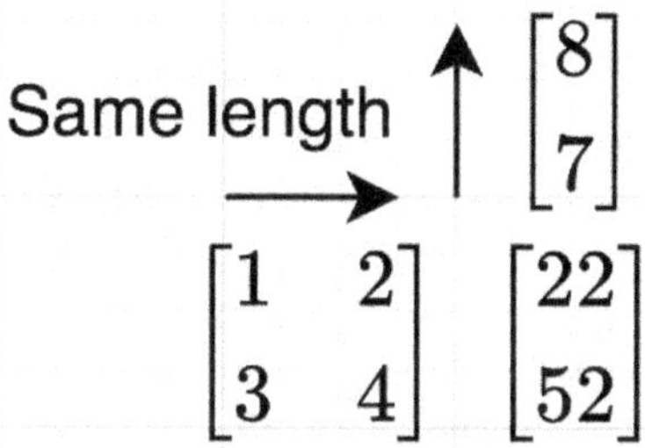

Now this extends for any arbitrary number of dimensions for any matrix, as long as the matrices are multipliable. We can see in the case of matrices and dot products how a dot product *precedes* matrix multiplication. Indeed matrix multiplication is essentially repeated dot products being performed, so one absolutely must know how to perform a dot product to multiply two matrices.

Property 2: Intrinsic Difficulty

We now come to the second key property of mathematics which is really just a fancy way of saying that it is a 'difficult/hard/tough' subject: *intrinsic difficulty*. But let's consider for a moment what is meant by *difficulty* - what do people really mean when they say that a subject is hard/difficult?

Within many different fields, there exist activities and pursuits which are definitely difficult, but their difficulty is mostly dependent on the skill required to perform and replicate an outcome.

Examples of this include playing the flight of the bumblebees on piano, making 5 consecutive three point shots in basketball or hitting the crossbar of a soccer goal, say 3 times in a row. These tasks are absolutely difficult to accomplish however performing them in one instance is nowhere near as difficult as recreating them within different parameters (game situations, pressure from a defender, live performance etc). In such tasks there already exists a base level of difficulty and the additional difficulty, or what *really* makes them difficult, is having to maintain the pattern-like behaviour in order to perform the same action again. If we are to overcome this kind of difficulty it can mostly be attributable to regular practice, usually in simulated environments that vary the parameters that can arise in execution. This is the reason why NBA players, cricketers and sportsmen in general practice incessantly -

to recreate a difficult task in an even more difficult pressurizing situation where their skills are really tested.

However this kind of difficulty by skill is very distinct from the difficulty inherent to mathematics and we can see why this is true in the following thought experiment. Suppose that we have an NBA basketballer and a ten year old. Now suppose you ask them to make 20 consecutive three pointers. Imagine that the NBA player then goes on to make the 20 consecutive three pointers and then the 10 year old *also* goes on to make the 20 three pointers. The ten year old did not have to go through a high school basketball program, the NCAA and a combine in order to achieve this accomplishment and technically the NBA player does not have to either. Now the NBA player has more *skill* (or at least he should) than the ten year old and can replicate this feat with greater likelihood, however if we are specifically referring to the *ability* to make twenty shots consecutively; it is still *possible* for a non-NBA (the ten year old) to accomplish the same feat. This is a key aspect of difficulty based on *skill*; that one does not need to pass any intermediary stages to accomplish the final task. The same case exists for a very difficult piece of music on piano for example; simply knowing the sequence of keys to be played is *technically enough* for a non-pianist to play an extremely difficult piece of music.

Now extending upon the same example, consider the difference between something like *playing* a difficult piece of music on piano and *creating the musical piece which is being played.* Creating the musical piece is a completely different territory which requires skill sets such as creativity, the recognition of instrumental compositions which 'sound good' together, and the ability to encapsulate specific emotions through sound (among others).

Another key skill here is the musician must know how to bring a composition, which purely exists in the mind of the musician, *into reality* and such skills are virtually unattainable *without* passing through some intermediate stage of musical learning. Whether it's having piano lessons, violin lessons, or instructions on musical theory. Now, since we know that it is a necessity to pass through these intermediate stages to *create* a musical piece (which we do not have to pass through when simply *playing* the musical piece), we can say that *creating* a piece of music is *intrinsically* more difficult than *playing* a piece of music. To reiterate in a simple statement:

Creating a musical piece is intrinsically more difficult than playing a musical piece since a practitioner has to go through more intermediate stages to create a piece than to play a piece (of music).

Similarly we can also quantify how *difficult* a field of learning (chemistry, biology, maths) is by measuring how many intermediate stages are required before being able to work with the most advanced aspects of the field.

Intrinsic difficulty and advanced concepts

Now that we have defined what intrinsic difficulty is, this gives us a way to measure how 'difficult' a concept in a field of study is based purely on the number of intermediate stages of learning a practitioner must be familiar with[43]. We can also quantify the number of preceding stages to understand a concept, for example:

From Mathematics:

Numbers → Addition → Multiplication → Exponentiation

So Exponentiation has three preceding stages.

From Biology:

Atoms → Molecules → Organelles → Cells → Tissue Formation

So Tissue Formation has four preceding stages.

From Physics:

Subatomic Particles → Atoms → Molecules → Forces → Motion

So Motion also has four preceding stages.

---

[43] I'm not the first one to do this. There's a reason your school curriculum considers calculus to be difficult and there's a reason it's taught in later schooling. All I'm saying is that we can measure how difficult a concept is in some objective way, aside from all the remarks students may be making about a subject.

With such a method we can begin to define which concepts in a field are more complex (advanced) or less complex (simple). The definitions for 'advanced' and 'simple' which we will use through the remainder of this text, follow quite directly from our discussions on intrinsic difficulty:

Advanced: requiring familiarity with many previous stages of learning (high intrinsic difficulty).

Simple: requiring familiarity with fewer stages of learning (low intrinsic difficulty).

Now the definitions have a degree of subjectivity in them - how many stages would constitute 'many previous stages of learning' for a concept to be advanced and how many stages constitute 'few stages' to label a concept simple. To remove this doubt, all I will say is that the reader can define this however they want to. Maybe 4 is the golden number. So in our examples above exponentiation in mathematics could be considered simple and motion in physics can be considered advanced. This choice is left up to the reader, the key is that there is some *distinction* between a concept being advanced or simple.

As is probably already known to the reader, calling a concept simple or advanced is not new. This is the reason why certain concepts within mathematics and the sciences are taught far later in any study design. Taking chemistry for example, we consider the periodic table to be advanced (which it is for the most part) because to understand the information it encodes requires *familiarity* with atomic numbers, orbitals, properties of elements etc.

This distinction between advanced and simple is important to realise because it gives some objective measure to the difficulty of a field. Thus when a practitioner is taught any of these concepts they should not feel disheartened if they *feel hard,* because they actually *are hard* in some objective sense.

Types of Intrinsic Difficulty

It's imperative within the sphere of mathematics and any intellectual domain to draw the distinction between tasks that are difficult to solve because of the number of computations (or steps) involved and tasks which have a level of conceptual difficulty. So when we class a computation as being difficult we specifically mean one of these two things:

1) The conceptual understanding required to perform a computation is high.
2) There are a high number of elementary steps required to obtain the final answer.

The first kind of difficulty we can give the name *difficulty by conception*. This is the case where a concept is actually difficult to understand or simply grasp the basis of, so to perform computations relating to such a concept (or to work with such a concept) will also be difficult. Within such a category we can include theorems such as general relativity, natural selection, blockchain, and a host of other theorems within the sciences and mathematics.

A point to emphasise here of importance is that these concepts *really are difficult* for *any practitioner* to grasp, especially if the concept has no connection to anything they have previously been exposed to. If one is exposed to various computer science concepts a field such as blockchain will be easier to understand, but to someone without a tertiary education or any kind of rudimentary background most will consider such concepts esoteric.

The second kind of difficulty is the difficulty inherent within a problem that simply requires a large number of steps to solve. It is usually these problems that every practitioner is examined on within schooling and is the main kind of difficulty they are exposed to within their educational journeys. This case we will refer to as *difficulty by computation*[44] and it can be summed up quite simply: one can observe that if a problem requires 100 steps to solve the main point of difficulty is characterised by the very fact that it requires *100 steps to solve* - regardless of how difficult each individual step may be.

[44] A point to note is that difficulty by computation also has an element of difficulty by conception within it as the steps required in attaining a final answer can themselves have an element of conceptual difficulty.

Difficulty by Computation

Difficulty by Computation is the main kind of difficulty a practitioner encounters in schooling. This is because mathematical problems, in general, are classified by their computational difficulty (the number of steps required to solve them). Thus it is then very natural to introduce new problems to practitioners, progressing through higher and higher year levels, based on increasing levels of computational difficulty. This subsequently gives the following fairly common graph which shows the different concepts taught in school against their computational difficulty which is followed by basically any practitioner following a western education system.

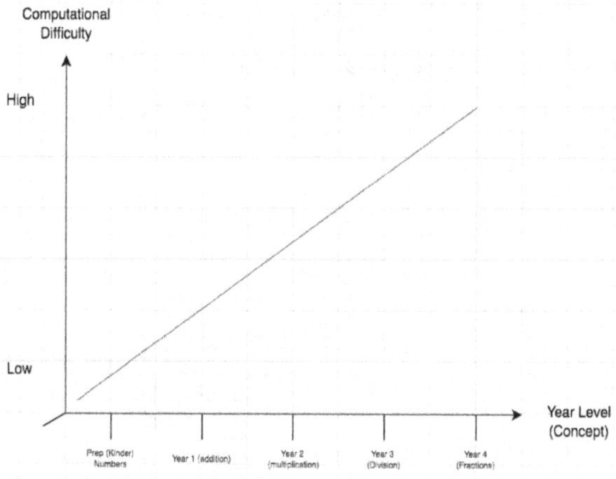

Computation Difficulty Graph: depicts the various concepts covered within schooling and how the computational difficulty of each concept progressively increases through each year level.

This graph can be surmised as follows. Practitioners will begin with the basics of addition and subtraction with numbers below 10, which they will usually learn by heart or their hands. Then adding two numbers together that are bigger than 10 requires more elementary steps (carrying over digits etc). Then adding even larger numbers that will require even more steps (albeit the same exact steps apply as those for numbers below 10) and around this same time we introduce multiplication.

If we speak purely in mathematical terms; upon the attainment of the conceptual understanding of a field, a problem will only be made more difficult by the number of elementary steps involved. In essence there is only so hard you can make certain concepts; adding two numbers, or finding the equation that defines a straight line can only be made so hard.

Difficulty by Conception

From our discussions above we have established that there is a difficulty that arises simply from a problem having a high number of steps required to solve it. Though each individual step may be simple, the sheer number of computations required to solve a problem is often the main source of difficulty.

However a question remains as to the difficulty that *does* exist in performing each individual computation. To be able to perform any *individual* computation requires *sufficient familiarity,* and well the thing is, some concepts are just plain difficult to work with and they require large degrees of exposure before a practitioner can really do anything.

When mathematics is taught it begins with concepts that are simple and the number of steps required in obtaining any answer is relatively low. For example addition, subtraction, multiplication are all treated as simple concepts because attaining any answer will usually (unless made long and convoluted) require fewer steps.

However there does exist another side to difficulty that lies in the disconnect between concepts that are taught and what we observe in the real world. Another reason why the exact same concepts (addition, subtraction and multiplication) are simple within mathematics is that they are easily visualised and do not need to be treated as abstract mathematical constructs the same way. matrices, vectors, tensors etc need to be.

When concepts begin to drift away from what we can visualise, or perceive through the senses in the real world, we immediately disconnect from familiarity with them. Often in such cases we try to create analogies within the real world to actually try to create some degree of familiarity.

An example of this is with derivatives and rates of change/speed whereby we try to explain the derivative in terms of the 'speed' of which a function is changing at any single point. However we must note that the concept of derivatives has nothing to do with speed or velocity in a literal sense; this is simply a tool for understanding derivatives.

Another example that can be used is the analogy between fractions and pizza. In any western schooling system I guarantee that fractions are first taught using pizza, since it conceptually ties in very well with fractions and it something the students will absolutely be familiar with. Fractions however do not care for pizzas at all! Again they are simply a tool we draw a parallel to within the field, to simplify familiarity attainment.

Abstraction and Intrinsic Difficulty

Continuing from our previous thought we can notice a connection between the relatability of a concept (whether one can perceive it in the real world) and its difficulty to grasp. In general we can say that the less relatable a concept is the harder it would be to have any familiarity with.

One question that immediately follows is then, what specifically about the low relatability of a concept increases its conceptual difficulty? What do we mean when we say that a concept is not relatable?

Well, if a concept cannot be related to anything a practitioner already knows then the practitioner will have no familiarity with the concept, and hence minimal numerical intuition. So to be able to perform any computations involving the concept becomes difficult for a practitioner. The second key aspect relevant to relatability is that there will inevitably be a number of steps and learning the practitioner must undergo to actually make the concept relatable.

If one is working within a purely abstract field (such as algebra or set theory) it is likely then that a practitioner simply has not performed enough computations with the objects of the field to be able to form some rudimentary level of numerical intuition. Without any intuition of the field, it then becomes much more difficult to understand and grasp. Coming back to our earlier example this is why it is much easier teaching a concept like fractions to practitioners through the pizza analogy because in some basic form, a practitioner would have already developed some intuition with fractions from having eaten pizza's (or at the very least seen a pizza) before!

Similarly a practitioner that has a deep interest in cars and motor vehicles would find the mathematical concepts of derivatives much easier to understand due the modelling of speed, velocity and acceleration through derivatives. In this way we can see that having some intuition formerly developed through real life experience will make an abstract concept much easier to work with for a practitioner.

Intrinsic Difficulty: The irrationality of the square root of two

The main concept within this section is the fact that mathematical problems generally have a high number of elementary steps to solve. Within the more theoretical domains of mathematics, the principle of *intrinsic difficulty is* even more visible within proofs, since proofs almost always entail a high number of elementary, necessary steps in order to solve.

Now it may be unclear to the reader at this point, what I mean when I say the word proof within the context of mathematics. Below are a few examples of statements, that would require a proof in order to solve:

1) Prove that the sum of two even numbers is also even.
2) In a right-angled triangle, the square of the hypotenuse is equal to the sum of the squares of the other two sides.
3) There are infinitely many prime numbers.

A practitioner can see how these kinds of problems have no immediate, obvious answer and that one must instead create an answer to solve them. The currently known proofs of the examples above, have a number of logical steps within them and we have thus classified these problems in terms of difficulty by the number of steps in their solution.

Now in order to highlight intrinsic difficulty in such problems let's look at an example. One of the most well known examples of a proof, is the proof of the irrationality of the square root of two. This proof also had wide scale implications on our understanding of numbers, since at the time of the Ancient Greeks people believed that you could basically write any number using other numbers in some way or another[45].

Take for example the number ¼. We call this a fraction and notice how it can be written using a 1 and a 4. Even some numbers which are infinite like 0.33333 recurring can be written as ⅓. So the Greeks had this idea that this would be the case for any number. You may be wondering why they would think such a thing, especially when they were blatantly wrong. Well, Greeks had a deep geometrical understanding of mathematics, and they would see numbers more from the perspective of lines and shapes then of discrete physical values.

If you were to present the question $x^2 = 16$ to a Greek and tasked them to find what x is they would see it as the following:

'What side length would a square need to have, so that the area of the square is 16 units?'

---

[45] To be more exact any number, p, could be expressed as the ratio of two integers a and b such that p = a/b.

If we were to perform such a question or computation ($x^2 = 16$), we likely would never interpret the question in such a manner, but if we look at this from the ancient Greek perspective then you'll probably try to find a square that has an area of 16 units. No wonder we have Pythagoras' theorem attributed to the Greeks, they were spending all the time drawing and measuring shapes where they noticed such a relation between a right angled triangle and its side lengths!

So the Greeks naturally believed then, that all numbers could be expressed by discrete values as we highlighted above. All of this changed though when Hippasus, a Greek philosopher, proved that the square root is irrational and his proof goes as such:

1. Assume that $\sqrt{2}$ is rational.

This means that we can express $\sqrt{2}$ as $\sqrt{2}$ = a/b where a and b are integers with no common factors (i.e in simplest form).

2. Write out $\sqrt{2}$ as it would be if it was rational.

   $\sqrt{2} = \frac{a}{b}$

3. Square both sides of the equation.

   $2 = \frac{a^2}{b^2}$

4. Multiply both sides by b^2.

   $2b^2 = a^2$

5. Now note that a^2 is even and since a^2 is even, a must also be even.[46]
6. Since we know that a is even, it can be written as $a = 2k$ for some integer k.
7. Substitute $a = 2k$ into the equation from above, $2b^2 = a^2$.

   This gives $2b^2 = 4k^2$.

   Which if we divide both sides by 2 gives: $b^2 = 2k^2$.
8. Now observe from the equation $b^2 = 2k^2$ that b^2 is even which means that b is also even (from the same logic as in step 5)
9. So we now observe that both a and b are even.
10. This contradicts our assumption that we can express $\sqrt{2} = \frac{a}{b}$, where a and b have no common factors.
11. Therefore our initial assumption that $\sqrt{2}$ is rational must also be false.
12. So we conclude that $\sqrt{2}$ is in fact rational.

[46] This actually requires a separate proof as well to verify, so we are actually skipping a number of elementary steps. Nonetheless, just from observation alone a practitioner can see that the following is true: a^2 is even if and only a is even.

Relation to Intrinsic Difficulty

There are few points to note with the proof above that may not be obvious to a practitioner. The first point is that no false statement can be derived from a true statement. This is why we first assumed that $\sqrt{2}$ is irrational, because we want to see if this assumption leads to any false statement. From the derivation of a false statement from any statement, we can then conclude that our initial assumption was false.

Now take note of how many *elementary* steps were required to reach the final result - a total of 12 absolutely required steps. Most problems a practitioner would encounter in high school would have at most 3 or four steps, which are simple and well defined. Many of the steps we took above are quite abstract and unobvious, a practitioner can never really be sure that their logical deductions are actually leading to the answer they may desire.

Thus with the proof of the irrationality of the square root of two we can see how the number of elementary steps required to solve the problem, entail a level of intrinsic difficulty to the problem itself. Additionally this proof is actually one of the most basic one encounters within mathematics, and by extension we can see that mathematical problems in general have a degree of intrinsic difficulty that simply does not exist in other fields.

Property 3: Interdependence

What's most interesting about mathematics and the way it has progressed throughout history is the relationship between problems and theorems. Throughout mathematical history what often happens is that, practitioners are usually faced with a problem which creates a kind of intellectual bottleneck and the solution births a whole new way of thinking about things. This follows a two step process where:

1) The problem is usually unsolved for a long period of time with numerous failed attempts made at its solution.
2) The solution is then so outlandish and unconventional that it stems a whole new way of thinking.

Now if we compare this to another field, say chemistry or physics, there are normally limitations to progression in the field by factors that are external to the field of study - normally by measuring instruments, testability, funding etc. We can see this in the case of the Large Hadron Collider (LHC) which was *created* to advance the field of physics[47] or in materials chemistry where one is limited in the development of new materials by the sheer volume of potential substances, compounds and access to them[48].

---

[47] The Large Hadron Collider (LHC) which was built in the period 1998-2008, was made specifically for the purpose of testing different theories of physics in relation to subatomic particles.

[48] Case in point is the famous invention of the light bulb filament by Thomas Edison (1847 - 1931), who is famously known to have tested around 6,000 different plant materials as potential filament sources for his light bulb before finding success with carbonized bamboo. This extensive experimentation was part of his broader effort, which involved thousands of trials and different materials to achieve a practical and long-lasting incandescent light bulb.

These cases represent instances where there is some kind of external requirement for the field to progress instead of the creation of new ways of thought for progression. In the case of problems that remain long unsolved in mathematics, these problems represent roadblocks to our understanding of mathematics. Often to actually create the solutions that overcome these roadblocks, requires directions of thought that are borrowed from other areas of mathematics or even philosophy.

The fact that a problem in an area of mathematics can be solved with concepts or ideas borrowed from a different area is actually a very interesting property unique to mathematics. What a practitioner will often find is that, this is often the case in mathematics - that the development of the mathematics comes from the merging of ideas and philosophies[49]. This helps us define the second key property of mathematics connected to the familiarity -

Interdependence: the property of a field where most concepts (if not all) are related to one another.

We can observe basic cases of interdependence through the following example. Any practitioner commencing the study of mathematics will almost always begin with numbers or shapes.

---

[49] Galois' proof of the insolvability of the quintic comes from the merger of group theory and field theory, Descartes in the creation of the Cartesian Plane which merges algebra and geometry and complex analysis is similar in the combination of its merger complex numbers and you guessed it - analysis - just to name a few examples.

These are the most fundamental building blocks that underlie all of mathematics and are related to virtually all other fields of mathematics in some way. Numbers are the most natural objects to commence with however even if study does begin with shapes, the shapes' distinctive properties are themselves described through numbers (triangle has 3 sides, square four sides etc). Then more advanced concepts are taught such as fractions, long division and multiplication then eventually moving to algebra. All of these concepts in some way depend on the base concepts of numbers or shapes and this extends even to very advanced areas such as matrices and topology.

Interdependence: The Cartesian Plane

Now you may not understand why I am asking this question, but you will in the coming pages:

'Why is Pythagoras' theorem such a big deal?'

Though this section is technically about the Cartesian Plane - the below question will be the beginning of our understanding of the significance of the Cartesian Plane.
Pythagoras' theorem is very simple and seemingly inconsequential when you first see it, just the statement below is all the information it contains:

'If a and b are the lengths of the shorter sides of a right angled triangle, the longer side is c, then we know that $c^2 = a^2 + b^2$ '

In fact Pythagoras' theorem predates the Cartesian plane by over 2000 years, which is absolutely astonishing if one really considers it. Why is this considered significant enough to be taught in schools even today?

To see why, let's play a game. It's called 'How far apart are the points?'.
Are you ready?

1)

2)

3)

Now it's pretty clear that the points are furthest apart in 2). but if you were to try and explain how far apart then you would need some kind of metric such as centimetres.

We could then rephrase our game instead to 'How many centimetres apart are the points?' instead of 'How far apart are the points?'.

So a practitioner can then use a ruler to physically measure how many centimetres apart each point is, but what if these points were not on a piece of paper. What if, say, there is a mountain in between each point and these points exist on Earth?

It is not practical in such a situation to rule 'over' the mountain, and any major obstacle in the path of the points will give an equivalent problem.

How then can one measure the distance between the two points, without really having to physically measure the distance between the points?

Well we can draw a right angled triangle of course, and measure the side lengths of that instead.

This is a wonderful idea, but where actually are the points?

Yes, you read that correctly, *where* are the points?

If a practitioner wants to measure the distance between the two points, using a right angled triangle, they need to have a point from which to measure distances (an origin point), and it is from this point that we can get the lengths of each side of the right angled triangle.

We can call one of the sides of the right angled triangle 'x' and the other 'y', and the point from which we measure distances is called the origin, which we can designate as having x and y equaling 0.

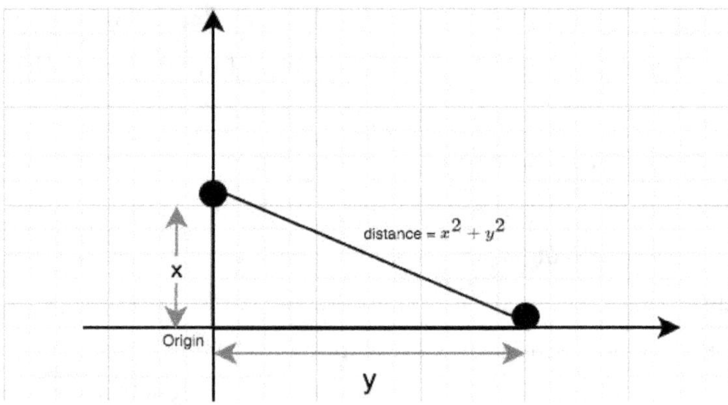

Then the end of each line (green and blue in the diagram) comes to represent how much 'x' or 'y' it has, in our case we are using centimetres as a metric, so the 'x' and 'y' lines will have a length also of 'x' centimetres and 'y' centimetres.

If we can imagine that we extend the lines for 'x' and 'y' infinitely we can then define a plane and using the same origin point we can also define different points. Similarly, we can also measure the distance of the origin to these points using a right angled triangle.

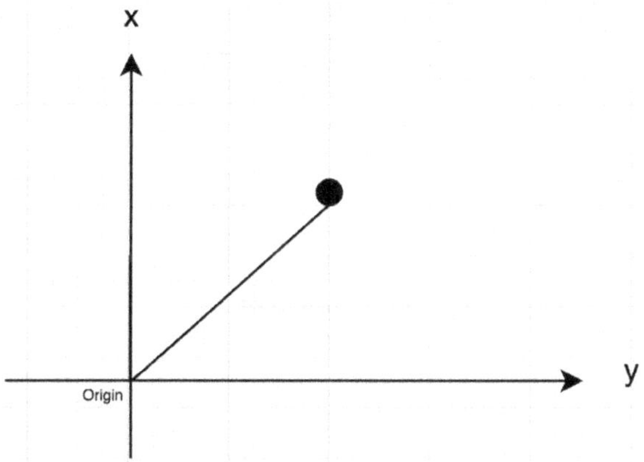

The case of the right angled triangle alone shows a deep connection between the Cartesian Plane and pythagoras' theorem and thus their *interdependence.*

Aside from distances

When we look at an equation such as $x + 10 = 15$, we can perform a subtraction to solve for what 'x' is to both sides of the equation. In maths terms if you add, subtract, multiply numbers etc, these are all called algebraic operations.

This is not the only way though that we can actually find x.
Using the cartesian plane that we defined earlier, if we have a defined unit on which to measure the axes, such as centimetres, then we can create points on the plane. Take for example the set of points below, and treat x and the first number and y as the second:

(1,2)
(2,3)
(3,4)
(5,6)
.
.
.
.

We can see the relation here between the x and y coordinate is that the y is one bigger than the x. So if we come back to the equation we have above, we can define the below points:

(1,11)

(2,12)

(3,13)

(4,14)

.

.

.

.

From here we can see that the relation between the x and y is that the y is 10 bigger than the x. The relation that we see here is exactly the same as the left side of the equation and the points above are ones that follow this relation.

So if we want to find the solution of the equation $x + 10 = 15$, then another way we can interpret this is, for what value of x on the above points do I get a y value of 15.

If we continue the relation as we listed for the above points:

(1,11)

(2,12)

(3,13)

(4,14)

(5,15) and bingo!

At the point (5,15) we see that the y value (the second number) is 10 bigger than x (the first number) and the solution of the equation is then x = 5. Using the points (1,11), (2,12) …. we can draw a line within the Cartesian Plane that defines all the points that follow the relation where its y value is 10 bigger than the x value. We can then visually check and mark at which x value the y value is then 10.

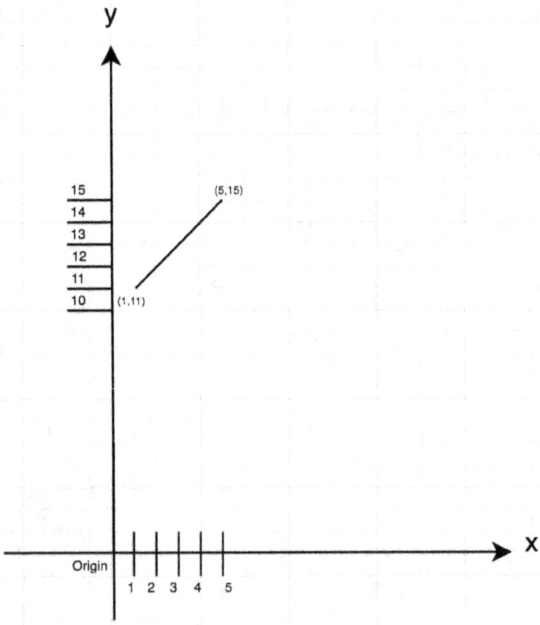

Now we can define lines through these relations but what of other geometric constructs. Circles clearly have a pattern to them, as do waves and vibrations - could we use this x-y value construct to try and define what their points are?

Well absolutely!
We won't detail these cases in depth but we can define virtually any shape using the Cartesian Plane (Geometry) and the x-y values (Algebra). Thus what we can see here is the interdependence and connective nature of the various fields within mathematics.

Summary: Numerical Intuition and Familiarity

What we discussed in the first chapter was the concept of numerical intuition, which is the inherent problem solving capability one develops through repeated computations/attempts on problems. Additionally, familiarity is how much continual exposure one has had to a concept.

What familiarity does is that it opens doors to computation and the development of numerical intuition. Once someone is sufficiently familiar with an area of maths they can work with objects within that field (numbers, fractions etc) and actually perform computations with them to then build their numerical intuition. This may seem very obvious with basic objects such as numbers and multiplication but extrapolate this out to matrices, proofs, groups then you can see why this is such an important acknowledgement. Having low familiarity i.e not understanding the rules for computations with these objects means there is no development of numerical intuition as one cannot perform the repeated computations/practice required.

So what is important to acknowledge here is that for one to develop their numerical intuition and perform well in any examination, they need to be sufficiently familiar with the relevant area. This is quite astounding. What I am saying is:

'To solve a problem you don't have to understand its components just be familiar with them and their rules'

This is especially important when reaching higher layers of abstraction in learning. At these points it can be incredibly difficult to find a real-world comparison to what you're learning (think economics, higher dimensions, many worlds). These kinds of concepts are incredibly difficult to work with, however, being sufficiently familiar with them allows a practitioner to progress in any of their related fields.

Part 5: End Notes

Gottfried Leibniz (1646 - 1716): "It is true that I consider mathematics as a study of quantity, but I find also that it is the science of order, which includes logic."

Remarks on Numerical Intuition

To begin this section, I would like to highlight the purpose and reason behind the construction of this work and how the ideas of this text first arose. Many of the ideas discussed within this text began through anecdotes and then through further contemplation, I was able to formalise them and present them as rigorous ideas within this book.

When I was in Year 7 and in fact in many of my years during schooling, I would notice many of my classmates lack in their speed and ability for mathematical calculation, though they had passed through the exact same system as I had. I never really understood why they couldn't perform the simple tasks they were asked by the teachers, thereafter often forming negative opinions of their teachers. I understood that they never did their homework, nonetheless if they *just* applied the instructions required for the computations they were assigned, then they would never fall behind. Yet they never did, and they continuously lagged further behind in class and began forming very negative (frankly self-destructive) opinions on their mathematical ability. However since mathematical ability is treated as a measure of intelligence in the present age, they would then think of themselves as frankly - *dumb.* Literally even saying such phrases in class:

'I'm too dumb to do this'
'I can't, I'm dumb'

‘If I wasn’t dumb I would have done X’

On a subconscious level, anyone that even minorly cares about their peers' self - opinions is then compelled to understand why they ever form such opinions in the first place and it’s never because of the actual difficulty of mathematics. Instead, what is the reason? The reason is a number of the ideas we have referenced in this book - essentially their lack of familiarity or performance of computation on any of the concepts they are taught. How this alone can spiral into such a negative worldview, is truly worth consideration.

When it comes to the relation of numerical intuition and familiarity to the real world there are always two key anecdotes that come to my mind and I am compelled to share them herein since they really encapsulate the purpose behind the creation of this text.

Anecdote 1: The Classmate

When I was in Year 7, one of my classmates was being troublesome in class (boisterous and noisy) so the teacher came and asked him to continue working on the questions that he was assigned. Though it was Year 7, we were working on multiplying fractions which is actually a primary school level task but at the beginning of year 7, at times, one can still be performing basic computations as part of assigned work (either homework or during the class).

Thus, I was watching as the teacher asked him to complete a question I had solved which was a simple fraction multiplication such as 2/7 * 8/11. The teacher was watching him go about solving the problem, and I was absolutely astonished when my classmate was tentative. He was stumbling in his words ('um', 'ah') and he was pretending to write to make it look like he knew the procedure. The teacher then came over and guided him; basically telling him the answer, 'The top gives 16 and the bottom 77'. The teacher then walked off asking another classmate if they needed assistance but the impression was left on me. To this day I'm still slightly shocked that he did not know how to solve the problem that - to use terminology we have created in this text - had minimal *intrinsic difficulty* (literally a two step computation).

I knew it had nothing to do with his intelligence level - the computation was tantamount to being able to follow an instruction, yet he did not know how to do it and I knew there was some reason at play. I could tell that his *familiarity* was low, since basic multiplication is essentially memorisation in most curriculums. So to perform multiplication, one can just memorise the answers but for this they need *continual exposure* to the answers otherwise no memorisation can occur.

Well then what was the outcome of the classmate?
He then stopped doing maths beyond Year 10, basically failing beyond Year 7. His inability to perform the computations initially, spiralled into a perceived inaptitude for mathematics hence leading to his decision to 'drop' the subject. If only he or his teachers were cognisant of some of the concepts we have discussed such an occurrence would be impossible.

Anecdote 2: The realisation

I was basically the same as any other person that did maths as an undergraduate subject. The only difference really was that my catalyst for pursuing mathematics came from two books: *Birth of a Theorem by Cedric Villani* and *So Good They Can't Ignore You* by Cal Newport and to a lesser extent *No Excuses* by Brian Tracy. *So Good They Can't Ignore You* and *No Excuses* contain messages which in hindsight are very obvious, but at the time revolutionary for anyone in high school, since it dispels so many of the contemporary narratives around work and the development of skills in the workforce.

Thus equipped with the backing of these texts, I made the choice to study mathematics - yet many of the teaching methods I encountered within university were exactly the same as high school.

What were these teaching methods?

To read lecture notes, watch the lectures and do practice problems in the classrooms/tutorials. Then eventually have examinations and so on, repeating this process each semester (as it likely happens in any university). One often conceives that the experience of tutelage in university would somehow be radically different from high school. But once you go there, you see that the teaching methods are really no different for mathematics and I anticipate that this is much the same for other subjects too.

The only real difference in university to high school, I found, was that the *time* for learning was greatly shortened. What a practitioner would spend probably a year on in high school (consecutively looking at the same thing or variations of concepts), a practitioner in university would do within a semester (about 12 weeks or less). Even despite such a greatly reduced period for learning, the methods of teaching in university and high school were basically the same so there must have been some basis behind it, I thought.

There were two reasons I thought that the teaching methods for mathematics at university could not be arbitrary:
1) Mathematicians are known for their deep thought on basically everything, so pedagogy was definitely an area they must have contemplated.
2) The teaching methods were the same even for proofs and more abstract areas of mathematics, which had virtually no computing or calculations involved..

As I continued on in university, classes where problems should have been otherwise easy to solve, started to become much harder and I felt I had lost my 'touch' within mathematics. I then noticed that I was in fact spending much less time revising and doing practice problems, thus much of these difficulties arose from a lack of *familiarity* with the course work.

Another core reason was that the nature of the problems presented were completely different to high school, where proving results was the main emphasis. Within university the professors and the pedagogical model was centred more around teaching mathematical reasoning from an axiomatic basis (simple statements from which you reason as opposed to performing computations) much like we introduced in earlier sections. However even before beginning to prove a result there was always *some computational problem* that was presented. Normally these preliminary problems would not be overly difficult and would have some relation to the topic of study. If we were trying to prove that two even numbers add to another even number then these preliminary problems would go something like this:

1) 'Add two even numbers'
2) Add two other even numbers again'
3) 'Now show that adding two even numbers always gives another even number'

So then what was the reason for the first two computations, I would question?

If the question we would actually be examined on was point 3 (which was purely theoretical) then what was the reason to perform two practical calculations as preliminary problems? We should be able to do the third point without the introduction of the first two right?

I had this thought during a number theory class, which deals purely with these kinds of problems. Around this same time I was also taking computer science classes, and I saw a relation between the creation of algorithms and the proofs we were studying in mathematics. My perspective was that an algorithm functioned as a proof, and a program for me was a proof by exhaustion which covers all potential cases of input through the various statements and logic encoded within the program. Much like a mathematical proof by exhaustion, one also covers simple cases within their program akin to points one and two, to test input of the program against expected output.

So within a practical and theoretical domain the same procedure of testing very basic inputs was performed prior to the creation of any algorithm or proof. So with these various elements within my mental frame, I was in the same class starting a problem set, and we began with two simple computational problems. While working on the problem, I asked myself 'what's the reason for performing these problems' to which I had an immediate answer:

'These computational problems build your intuition'

This coincided with the fact that I was losing my mathematical touch in the case of new concepts being taught, and I could clearly see the reason why. My intuition - numerical intuition - in the concepts being taught was very low, because of the shorter time frame and lower intensity of practice I was having when compared with high school. The level of continual exposure to a concept was very low hence creating low familiarity. Furthermore, because of precedence and interdependence, lagging behind in the initial phases of a unit meant that I would simply fall further behind and struggle more in classes later in the semester. All of this would be negated coming into the exam period where there was a phase of continual exposure and numerical intuition development.

Additionally problems at such levels have a high degree of intrinsic difficulty since they can be incredibly abstract, having sometimes no relatability to the real world. Thus in such areas it is even more imperative to perform simple computations to develop the intuition required to tackle such problems.

Conclusive Note

Thus we now reach the end of this first text on numerical intuition. Within this excerpt we covered what numerical intuition is, which to mention the definition one final time:

Numerical intuition: the inherent problem solving ability one develops from repeated computations and attempts on problems.

We started this text by breaking down the definition of numerical intuition, analysing the meaning of each individual component of the definition looking at the meaning of inherent, problem solving ability and repeated computations. In order to discuss numerical intuition we covered the related concept of *sufficient familiarity,* which is the minimum level of familiarity required to begin performing the repeated computations for numerical intuition development. To highlight the existence of the sufficient familiarity case we analysed the lives of Darwin and Mendeleev, whose scientific careers display cases of the usage of sufficient familiarity for discoveries.

In order to see why numerical intuition is a viable model for mathematical learning, we analysed some of the key properties of mathematics which included *precedence, intrinsic difficulty* and *interdependence* - looking at the cases of Einstein, Galois and Wiles among others to demonstrate the relevance of these properties to their mathematical/physical discoveries.

Again I hope the reader has taken an idea or concept from this text they can implement within their own learning, or at the very least, enjoyed reading this text. In the second component of numerical intuition we will look at *understanding* and how numerical intuition helps us to form an understanding of a concept.

Reference List

[1] **Duffy, C., & Branley, A. (2025, July 29).** *NAPLAN results again show one-third of students aren't meeting literacy and numeracy expectations amid calls for urgent change.* ABC News. Retrieved from https://www.abc.net.au/news/2025-07-30/naplan-national-testing-results/105581218

[2] Wittrock, Christian. (2015). Reembedding Lean: The Japanese Cultural and Religious Context of a World Changing Management Concept. International Journal of Sociology. 45. 95-111. 10.1080/00207659.2015.1061852

[3] Gilmore Gerard and Tausch-Pebody Gudrun 2022. The 1919 eclipse results that verified general relativity and their later detractors: a story re-told *Notes* Rec.**76** 155–180 http://doi.org/10.1098/rsnr.2020.0040

[4] Roy, A. Science and the Large Hadron Collider: a probe into instrumentation, periodization and classification. Dialect Anthropol 36, 291–316 (2012). https://doi.org/10.1007/s10624-012-9278-6

[5] Deborah Peres Menezes, *A Neutron Star is born,* Universe 2021, 7(8), 267; https://doi.org/10.3390/universe7080267

[6] Goddi, C., Crew, G., Impellizzeri, V., Martí-Vidal, I., Matthews, L., Messias, H., Rottmann, H., Alef, W., Blackburn, L., Bronzwaer, T., Chan, C., Davelaar, J., Deane, R., Dexter, J., Doeleman, S., Falcke, H., Fish, V., Fraga-Encinas, R., Fromm, C., Herrero-Illana, R., Issaoun, S., James, D., Janssen, M., Kramer, M., Krichbaum, T., De Laurentis, M., Liuzzo, E., Mizuno, Y., Moscibrodzka, M., Natarajan, I., Porth, O., Rezzolla, L., Rygl, K., Roelofs, F., Ros, E., Roy, A., Shao, L., van Langevelde, H., van Bemmel, I., Tilanus, R., Torne, P., Wielgus, M., Younsi, Z., Zensus, J., The Event Horizon Telescope collaboration; First M87 Event Horizon Telescope Results and the Role of ALMA. The Messenger 177 (September 2019): 25–35. https://doi.org/10.18727/0722-6691/5150

[7] Observation of Gravitational Waves from a Binary Black Hole Merger. Phys. Rev. Lett. 116, 061102 – Published 11 February, 2016, https://doi.org/10.1103/PhysRevLett.116.061102.

[8] **Antognazza, Maria Rosa**. Leibniz: An Intellectual Biography. Cambridge University Press, 2009.

[9] Huistra, Noortje,. (2019). Funding Bodies and Late Modern Science. International Journal for History, Culture and Modernity. 7. 10.18352/hcm.584.

[10] Singh, Simon, and Kenneth A. Ribet. "Fermat's Last Stand." Scientific American, vol. 277, no. 5, 1997, pp. 68–73. JSTOR, http://www.jstor.org/stable/24996003

[11] Dana Mackenzie ,The Poincaré Conjecture--Proved. Science **314**, 1848-1849 (2006) . DOI:10.1126/science.314.5807.1848

[12] Vali Siadat, M., & Tholen, A. (2020). Omar Khayyam: Geometric Algebra and Cubic Equations. Math Horizons, 28(1), 12–15. https://doi.org/10.1080/10724117.2020.1770495

[13] Mike Thelwall, Nabeil Maflahi; Research coauthorship 1900–2020: Continuous, universal, and ongoing expansion. Quantitative Science Studies 2022; 3 (2): 331–344. doi: https://doi.org/10.1162/qss_a_00188

[14] Kanigel, R. (1991). *The man who knew infinity: A life of the genius Ramanujan* (Chapter 7: "The English Chill", p. 256). New York: Charles Scribner's Sons.

[15] **Browne, Janet**. Charles Darwin: The Power of Place. Princeton University Press, 2002.

[16] **Buxton, D. M. Mendeleev**: The Story of the Periodic Table. Cambridge University Press, 1985.

[17] **Gleick, James**. Isaac Newton. Vintage Books, 2003.

[18] Hamel, J., "Kepler, Galileo, the telescope and its consequences", Astronomische Nachrichten, vol. 330, no. 6, p. 526, 2009. doi:10.1002/asna.200911211.

Images List

Page 12: Niels Bohr

Source: Public Domain,

https://commons.wikimedia.org/w/index.php?curid=288274

Page 25: John Von Neumann

Source: LANL

http://www.lanl.gov/history/atomicbomb/images/NeumannL.GIF3

Page 28: Andrew Wiles

Source: Copyright C. J. Mozzochi, Princeton N.J, Attribution,

https://commons.wikimedia.org/w/index.php?curid=2635913

Page 29: Grigori Perelmann

Source: George M. Bergman -

https://opc.mfo.de/detail?photo_id=12890

Page 43: John Conway

Source: Thane Plambeck -

"https://www.flickr.com/photos/thane/20366806/", CC BY 2.0,

https://commons.wikimedia.org/w/index.php?curid=13076802

Page 74: Charles Darwin

Source: Julia Margaret Cameron / Adam Cuerden - Alfred Steiglitz Collection reference 1949.881, Art Institute of Chicago, Public Domain, https://commons.wikimedia.org/w/index.php?curid=131628151

Page 86: Dmitri Mendeleev

Source: Public Domain,

https://commons.wikimedia.org/w/index.php?curid=32692107

Page 98: Dmitri Mendeleev

Source: Godfrey Kneller - File:Portrait of Sir Isaac Newton, 1689.jpg from https://exhibitions.lib.cam.ac.uk/linesofthought/artifacts, Public Domain,

https://commons.wikimedia.org/w/index.php?curid=132521185

Page 112: Rene Descartes

Source: Frans Hals - André Hatala [e.a.] (1997) De eeuw van Rembrandt, Bruxelles: Crédit communal de Belgique, ISBN 2-908388-32-4., Public Domain,

https://commons.wikimedia.org/w/index.php?curid=2774313

Page 160: Gottfried Leibniz

Source: Christoph Bernhard Francke - Herzog Anton Ulrich-Museum, online, Public Domain,

https://commons.wikimedia.org/w/index.php?curid=53159699

www.ingramcontent.com/pod-product-compliance
Lightning Source LLC
La Vergne TN
LVHW022218200926
848481LV00006B/2021